The Ambonese Herbal

The Ambonese Herbal

Being a Description of the Most Noteworthy
Trees, Shrubs, Herbs, Land- and Water-Plants
Which Are Found in Amboina and the Surrounding Islands
According to their Shape, Various Names,
Cultivation, and Use: Together with
Several Insects and Animals.

Georgius Everhardus Rumphius

Translated, annotated, and with an introduction by

E.M. Beekman

Volume Six

Index of Common and Scientific Names

Yale UNIVERSITY PRESS *&* National Tropical Botanical Garden
New Haven and London

Yale University Press books may be purchased in quantity for educational, business, or promotional use.
For information, please e-mail sales.press@yale.edu (U.S. office) or sales@yaleup.co.uk (U.K. office).

Printed in the United States of America.

Library of Congress Cataloging-in-Publication Data

Rumpf, Georg Eberhard, 1627-1702.

[Amboinsche kruidboek. English]

The Ambonese herbal : being a description of the most noteworthy trees, shrubs, herbs, land-
and water-plants which are found in Amboina and the surrounding islands according to their shape,
various names, cultivation, and use : together with several insects and animals : for the most part
with the figures pertaining to them : all gathered with much trouble and diligence over many years
and described in twelve books / by Georgius Everhardus Rumphius ; translated by E.M. Beekman.

p.   cm.

Includes index.

ISBN 978-0-300-15376-7 (v.1-6 : alk. paper) — ISBN 978-0-300-15370-5 (v.1 : alk. paper) —
ISBN 978-0-300-15371-2 (v.2 : alk. paper) — ISBN 978-0-300-15372-9 (v.3 : alk. paper) —
ISBN 978-0-300-15373-6 (v.4 : alk. paper) — ISBN 978-0-300-15374-3 (v.5 : alk. paper) —
ISBN 978-0-300-15375-0 (v.6 : alk. paper)

1. Botany—Indonesia—Ambon Island. 2. Herbals—Indonesia—Ambon Island.
3. Botany—Pre-Linnean works. I. Beekman, E.M., 1939–2008. II. Title.

QK367.R8213   2010
581.9598'52—dc22

A catalogue record for this book is available from the British Library.

This paper meets the requirements of ANSI/NISO Z39.48-1992 (Permanence of Paper).

10   9   8   7   6   5   4   3   2   1

Donors and Benefactors

Publication of *The Ambonese Herbal* has been made possible by the generous support of the following benefactors:

    Ms. Leslie Clarke

    Mr. Douglas McBryde Kinney

    National Tropical Botanical Garden, Kauai, Hawaii

Research grants to E.M. Beekman for development and preparation of *The Ambonese Herbal* were awarded by

    John Simon Guggenheim Memorial Foundation

    The Richard Lounsbery Foundation

    The Rockefeller Foundation

    Prince Bernhard Cultural Foundation

    Foundation for the Translation of Dutch Literary Works

Scanning and preparation of the plates was provided by

    Missouri Botanical Garden, St. Louis, Missouri

Institutions and foundations that also aided completion of *The Ambonese Herbal*:

    National Herbarium of the Netherlands, Leiden, Netherlands

    Naturalis: National Museum of Natural History, Leiden, Netherlands

    The Kampong, National Tropical Botanical Garden, Miami, Florida

    The Montgomery Botanical Center, Coral Gables, Florida

    Fairchild Tropical Botanic Garden, Coral Gables, Florida

    Rijksmuseum, Amsterdam, Netherlands

    The Mauritshuis, The Hague, Netherlands

    The University of Massachusetts at Amherst, Massachusetts

    National Education Foundation for Moluccans, Utrecht, Netherlands

Index of Common and Scientific Names

This sixth volume of *The Ambonese Herbal* is a combined index of the common and scientific names of plants in the translated text, notations, and introduction by E.M. Beekman. Entries include the volume number followed by a colon and the page number where the entry is found. For example, the first entry in the index, "Aäl, 1:277" is found in volume 1 on page 277. The second entry, "Aal tree, 3:151" is found in volume 3 on page 151. Entries in the index that appear in footnotes are preceded by an "*n*" followed by the footnote number. For example, "*Abroma fastuosa*, 3:241*n1*" is found in volume 3 on page 241 in footnote 1.

A

Aäl, 1:277

Aal tree, 3:151

Aal-klim, 3:469

Abanus, 2:412

Abatj, 1:491

Abbattiach, 4:429, 4:431

ABC daria, 5:278, 5:279

Abel-mosch, 3:244

Ahelmoschus
    mindanaensis, 3:244*n11*
    moschatus, 1:431*n9*, 1:432*n1*,
        2:304*n17*, 2:527*n6*, 3:243*n1*,
        4:62*n32*

Abhu maün, 4:405

Abies, 5:381
    marina, 5:377, 5:405

Abroma fastuosa, 3:241*n1*

Abrus
    alpini, 3:567, 3:568
    arbor, 3:567
    frutex, 3:567
    precatorius, 3:230*n10*, 3:565*n1*,
        4:270*n31*

Abrus shrub, 3:565, 3:566, 3:567–
    3:568

Abubo, 4:358

Abutilon, 3:134*n6*, 4:192
    agreste, 3:228, 3:229–3:231
        sive montanum, 3:230
    albescens, 3:231*n18*
    domesticum, 3:225, 3:226, 3:227
    hirsutum, 3:225, 3:226, 3:227
    hirtum, 2:334*n12*
    indicum, 2:334*n12*, 3:227,
        3:229*n1*, 4:192*n17*
    laeve, 3:228, 3:229–3:231
    litoreum, 3:231, 3:231*n18*
    sonneratianum, 3:231*n16*
    sonneratium, 1:31

Acacia, 1:234*n31*, 3:267
    aegyptiaca, 4:270
    arabica, 4:270*n40*
    Ceylonese, 3:272
    holosericea, 2:637*n1*
    mangium, 2:637*n1*
    tinctoria, 3:272

Acalypha
    amentacea, 3:321*n1*
    boehmerioides, 5:91*n1*
    hispida, 3:319*n1*

Acanna, 2:631

Acanthus, 4:199
    ebracteatus, 2:403*n27*, 5:311*n2*
    ilcifolius, 5:311*n2*
    volubilis, 5:311*n4*

Acaricoba, 4:525

Accar
  binassa, 3:359, *4:520, 4:521–*
    4:522
  cussu, 5:150
  mera, *5:474, 5:475*
  pamali, 3:553
  pouloron, *4:559, 4:560*
  ticos, 5:523
Accar Lampau root, 2:193
Accarbaar, 5:357, 5:359
  abu abu, 5:407
  batu, 5:401
  boa sagu, 5:402
  caju, 5:401, 5:407
    alus, 5:405
  djinka, 5:421
  gabba gabba, 5:402
  kipas, 5:376
  lacki lacki, 5:369*n2*, 5:371
  mera, 5:421
  puti, 5:401
  rompot, 5:379
  ruttu ruttu, 5:381
Accarbarium
  album, 5:411, *5:412*, 5:413–5:417
    littoreum, 5:414
    pelagium, 5:414
    verrucosum, 5:419
  cinereum
    ericae forma, 5:401
    ramosum, 5:401–5:402
    unicaule, 5:401
  ericae forma, 5:405
  montanum, 5:399
  nigrum, 5:361, *5:362*, 5:363–
    5:368
  pelagium, 5:410, 5:417
  rubrum, 5:420, 5:421–5:422
  unicaule, 5:369, *5:370*, 5:371–
    5:373
    album, 5:409–5:410
Acebar, 4:221
Acer campestre, 2:265*n7*, 4:447*n4*
Aceratum oppositifolium, 3:431*n1*
Acetabulum marinum, 5:345–5:347
Acetosa arbor, 2:513
Acetosella agrestis, 3:245
Achasma megalocheilos, 5:273*n22*

Acherontia styx, 3:365*n2*
Achyranthes aspera, 5:49*n4*, 5:80*n11*
Ackar pouloron, 4:555
Acmella paniculata, 5:279*n1*
Aconitum, 4:295*n2*
  lycoctonum, 4:296*n3*
Acorus, 4:33*n7*, 4:63, 5:80*n18*
  calamus, 3:258*n17*, 3:463*n3*,
    4:48*n6*, 4:63*n44*, 4:81, *4:82*,
    4:83–4:84, 4:138*n30*, 4:219*n9*,
    4:239*n11*, 4:323*n31*, 4:513*n9*,
    5:13*n9*, 5:38*n9*, 5:51*n18*,
    5:80*n12*, 5:353*n3*, 5:51*n7*,
    5:555*n11*
  marinus, *5:336*, 5:353–5:354
  water, 4:81
Acropora arbuscula, 5:429*n5*
Acrostichum aureum, 5:507*n1*
Actinodaphne
  moluccana, 3:71*n1*
  rumphii, 3:71*n5*
Actinorhytis calapparia, 1:223*n1*,
  1:226*n29*
Adam's Apples, 2:142, 4:6, 4:13
Adamaram, 1:467
Adas adassan, 4:249
Adder's tongue, 4:367*n14*, 5:147*n2*,
  5:295*n1*
Adecca, 1:225
Adenanthera
  microsperma, 3:84*n6*
  pavonina, 2:310*n14*, 3:83*n1*,
    3:563*n3*, 3:567*n6*
Adenosma capitatum, 5:555*n1*
Adenostemma lavenia, 5:65*n2*,
  5:72*n10*, 5:80*n15*
Adhar, 4:86
Adianthum, 5:138, 5:139, 5:140*n7*
  capillus-veneris, 5:364*n15*
  nigrum, 5:140
  opacum, 5:145*n1*
  volubile, 1:214
  volubile polypoides, 5:141, 5:142,
    5:143
Adina fagifolia, 2:479*n1*

Adinauclea fagifolia, 2:120*n10*

Adju Java, 2:505

Adpendix
  arborum, 4:570, 4:571–4:572
  cuscuaria, 4:578, 4:579
  erecta, 4:574, 4:577
  gumuti, 4:572
  laciniata, 4:578, 4:580, 4:581–
    4:582
  porcellanica, 4:573, 4:574, 4:575

Adrac, 4:51

Adrae, 4:55

Adulterina, 5:110, 5:111–5:112

Adultery plant, 5:110, 5:111–5:112

Advent wood, 2:56

Aegiceras
  corniculatum, 2:625*n1,* 2:639*n1,*
    2:639*n6*
  floridum, 2:582*n5,* 2:627*n5,*
    2:631*n6,* 2:641*n1*

Aegle marmelos, 1:505*n1,* 1:509*n1,*
  2:162*n9,* 2:198*n9*

Affu, 3:127

Aforo, 4:358

Afternoon flower, 4:192*n13,*
  4:247*n6*

Agallochum, 3:25, 3:501*n17*
  bastard, 3:332

Agallochum, 2:47–2:51, 2:53–2:62,
  2:80, 2:361, 2:559
  album, 2:62
  albumericum, 2:60
  crudum, 2:59
  indum, 2:60
  mondunum, 2:60
  primarium, 2:47, 2:74
  secundarium, 2:47, 2:53–2:62,
    2:363
  seificum, 2:60
  silvestre, 2:60, 2:61
  simplex, 2:62
  spurium, 2:61, 2:88

Agar agar, 5:346, 5:433
  carang, 5:339
  cupan, 5:345

Agaricaceae, 5:249*n1*

Agarum
  bracteatum, 5:345, 5:346
  coralloides, 5:339–5:340, 5:348
  corticosum, 5:346
  lactucarium, 5:346
  secumdum, 5:345, 5:346

Agasto, 1:491

Agathis alba, 2:12*n8,* 2:263*n1,*
  2:269*n1,* 2:637*n2,* 3:338*n13,*
  4:535*n17*

Agatj, 1:491

Agave, 3:453*n6, n14*
  cantala, 4:224*n22*

Agel wood, 2:52, 2:53–2:62,
  2:314*n7,* 2:316, 2:361, 2:366
  bastard, 2:61–2:62, 2:360, 2:365
  conterfeit, 2:59–2:60

Agel-hout, 2:54

Agelwood, 3:270, 3:271, 3:501,
  3:523, 3:559, 4:63
  scented, 4:223

Ageratum conyzoides, 5:65*n5*

Agir, 4:83

Agis, 4:188

Aglaia silvestris, 1:425*n1,* 3:275*n15,*
  5:473*n610,* 5:561*n2*

Aglaonema oblongifolium, 2:115*n2,*
  4:283*n1,* 4:355*n6,* 4:577*n1*

Aglaophotis, 5:552

Aglata adorata, 3:514*n15*

Agnus castus, 3:255, 3:261, 3:373

Agrestis minor, 5:19–5:20

Agrimonia, 3:247, 5:71*n2,* 5:279
  indica, 5:279
  molucca, 1:65, 5:68, 5:71–5:72

Agrimony, 3:247*n7*
  hemp, 3:247*n6*

Agrostis spica venti, 5:19*n7*

Ahaain, 5:215, 5:219
  malona, 5:219

Ahalim, 4:223

Ahamahu, 1:502

Ahiar, 4:357

Ahoë ila, 4:401

Ahuellu, 2:428

Ahuhe mamehun, 4:405

Ahuho, 4:349, 4:359

Aiger-Indi, 1:196

Ailanthus moluccana, 3:149*n1*

Aile, 5:140, 5:151

Airi, 1:284

Ajer puti, 1:290

Ajoccol, 4:521

Alaf elgemali, 4:86

Alalun, 4:119

Alang-alang, 3:637, 5:33*n1*

Alaun
   pinan, 4:575

Albahaca, 4:207

Albatheca, 4:432

Albizia, 3:90*n8*, 3:119*n9*
   falcata, 2:497*n14*, 3:87*n1*,
      3:135*n3*, 3:266*n10*
   procera, 2:129*n7*, 2:495*n1*,
      3:135*n2*
   retusa, 2:497*n16*, 3:135*n1*,
      3:277*n27*
   saponaria, 2:200*n6*, 3:413*n1*

Alcali plant, 4:136

Alcanna, *3:248*, 3:249–3:254, 3:252,
   4:136
   aquosa, 4:196

Alcea
   acida spinosa, *3:245*, *3:246*, 3:247
   Americana, 3:247
   veneta, 3:247

Alchaemenidos, 5:551

Alchanna, 3:250

Alchimilla, 4:525

Alcyonium, 5:421*n2*
   primum peregrinum, 5:422

Alea
   pade, 4:55
   utan, 4:41

Alegria, 4:119

Aleurites
   fordii, 2:271*n1*
   moluccana, 1:278*n28*, 2:271*n1*
   moluccanus, 4:135*n8*

Alexicacum, 3:543, 3:553, 3:554

Alexiteria, 5:51

Alga
   bracteolata, 5:346
   coralloides, 5:171, 5:339–5:340,
      5:341, *5:348*, 5:358, 5:358*n13*,
      5:433
   esculenta, 5:340

Algodon, 3:234

Alhalla, 3:251

Alhenna, 3:251

Alkaroa, 3:340

Alliaria, *2:118*, 2:119–2:120

Alligura banga, 3:327

Allium porrum, 5:3*n2*

Allophylus timorensis, 1:509*n8*,
   2:286*n7*

Allpenia galanga, 3:313*n2*

Almond tree, 1:467

Alocasia
   indica, 4:277*n4*
   longiloba, 4:281*n2*, 4:572*n3*
   macrorrhizos, 3:378*n17*,
      4:137*n19*, 4:138*n34*, 4:277*n1*,
      4:281*n1*, 5:301*n8*

Aloe, 4:221, 4:222, 4:223–4:225
   Americanam, 4:223, 4:224, 4:225
   ferox, 4:221*n1*
   vera, 2:87*n13*, 3:541*n5*, 4:221*n1*,
      5:157*n4*

Aloës, 2:47–2:51

Alopecoris, 5:34

Alosna de batao, 4:201

Alphitonia moluccana, 5:481*n1*,
   5:481*n3*

Alpinia, 2:310*n13*, 5:5*n13*, 5:267*n3*,
   5:273*n20*
   galanga, 2:454*n6*, 2:561*n1*,
      3:332*n19*, 3:372*n14*, 4:30*n17*,
      4:31*n1*, 4:48*n6*, 4:57*n1*,
      4:63*n46*, 4:95*n14*, 4:138*n32*,
      4:323*n31*, 4:571*n2*, 5:11*n5*,
      5:261*n2*, 5:263*n17*, 5:267*n4*,
      5:317*n11*
   gigantea, 5:273*n11*, 5:559*n6*

malaccensis, 2:461*n3*, 3:215*n2*, 4:77*n1*, 5:267*n2*
nobilis, 4:77*n1*
nutans, 4:29*n11*, 5:271*n1*, 5:273
padacana, 5:273*n20*
regia, 5:273*n11*
uviformis, 4:78*n6*, 5:267*n1*

Alsine, 5:43
indica, 5:40, 5:43–5:45, 5:175

Alstonia
acuminata, 1:277*n18*, 2:454*n1*
scholaris, 2:15*n25*, 2:373*n1*, 2:391*n4*, 2:406*n13*, 2:434*n20*, 2:454*n2*, 3:173*n3*, 3:320*n7*, 3:520*n11*, 3:521*n2*, 3:543*n20*, 4:300*n17*, 5:105*n9*, 5:109*n17*, 5:521*n3*
subsessilis, 1:277*n18*, 2:375*n16*, 2:454*n1*

Alternanthera sessilis, 5:69*n1*

Althaea, 2:173, 2:331, 3:217*n2*, 3:243
flore luteo, 3:227
horaria, 3:222, 3:223–3:224
italica, 3:227
officinalis, 3:217*n2*

Altingia excelsa, 2:83*n1*, 2:88*n33*

Alum tree, 3:57, *3:58*

Alumen plumbosum, 4:419, 4:419*n6*

Alwag, 4:83

Alwah, 4:83

Alya, 4:51

Alyxia, 2:577*n6*, 3:47*n3*, 3:491*n5*, 5:479*n6*
laurina, 1:363*n23*, 2:183*n12*, 3:41*n7*, 3:372*n12*, 3:521*n1*, 3:615*n6*, 3:622*n7*, 4:62*n31*, 5:126*n17*, 5:471*n15*, 5:479*n6*, 5:571*n2*
Gaudich, 3:417*n2*

Amaëka, 3:8

Amaena maesta, 5:283, *5:284*

Amakin Abbal, 3:109

Amale, 3:60

Amara
indica, 4:447, *4:448*, 4:449

litorea, 3:595
silvestris, *4:450*, 4:451, 4:454

Amaranthus, 1:481*n1*, 3:295*n4*, *3:318*, 3:319, 3:560*n24*, 4:160, 4:163, 5:545*n6*
Ceylonese, 5:53
gangeticus, 4:159*n7*
globosus, 4:249
gracilis, 5:65*n4*, 5:69*n4*, 5:83*n4*
Japonicus, 4:115*n2*
lividus, 4:159*n7*
saxeus, *5:436*, 5:437–5:438, 5:444
Sicilian, 5:53
spicatus, 4:168, 5:577
spinosus, 3:182*n15*, 4:161*n1*, 5:511*n3*
tricolor, 2:406*n9*, 4:157*n1*, 4:157*n3*, 4:157*n5*, 4:157*n6*, 4:168, 4:168*n8*
versicolor, 4:168
viridis, 2:577*n4*, 4:157*n1*, 4:157*n3*, 4:157*n5*, 4:191*n4*, 4:521*n3*, 5:44*n8*, 5:49*n2*, 5:95*n2*, 5:115*n5*, 5:119*n2*, 5:257*n3*

Amare, 3:113

Amargosa, 4:447

Amaryllidaceae, 3:269*n3*

Amaryllis, 4:275

Amassi, 5:473

Amba, 1:327

Amba-paja, 1:418

Ambalam, 1:441

Ambalo, 3:251

Ambela, 5:529

Ambergunong, 2:81

Ambig, 1:327

Amboinze boontjes, 4:401

Ambonese tea, 5:86

Ambra Montana, 2:237

Amelaun
Abbal, 2:521
tallan, 3:582

Amelikaan, 3:336

Amenduinas, 4:474

Amianthus, 4:419

Amica nocturna, 4:240, 4:241–4:242

Ammelaun, 4:320
  abber mamare, 4:333
  albar, 4:329
  halat, 4:329
  mäate, 3:549
  matta rulalan, 4:332
  nitu, 3:546
  tallan
    laun maun, 3:547
    puti, 3:546
  tallan mahina, 4:333
  une, 4:331

Ammomum cardamomum, 1:234n28

Amoenitatum exoticarum, 2:385n19

Amomum, 4:77n4
  aculeatum, 5:265n1, 5:266n7
  cardamomum, 4:43n1, 4:365n3,
    5:220n10, 5:262n6
  gymnocarpum, 5:266n7
  magnificum, 5:273n17
  maximum, 4:45n7
  roseum, 5:219n6, 5:265n6
  rumphii, 1:276n12, 1:279n32,
    3:461n2, 5:261n1

Amoora amboinensis, 1:428n10

Amorphophallus
  campanulatis, 4:301n1
  paeonifolius, 4:301n1, 4:305n1,
    4:305n3, 4:358n6

Amourette, 5:23, 5:23n5, 5:24

Ampac Tree
  broad-leaved, 2:281, 2:282, 2:283
  small-leaved, 2:284, 2:285–2:286

Ampacus, 2:281, 2:282, 2:283
  angustifolia, 2:284, 2:285–2:286,
    4:332
  litorea prima, 2:286

Ampas ampas, 4:477, 4:479, 4:481

Ampatsjan, 1:333

Ampelaas, 3:404, 3:405–3:406

Ampellas, 3:490

Ampelocissus arachnoidea, 4:519n1,
  4:561n10

Ampetu tana, 4:529

Ampiri Tree, 2:273

Ampulat pulat, 5:113

Amudium, 4:219

Amulen, 4:470
  lanit, 4:251
  lauaha, 4:470

Amusa, 4:6

Amvalli, 5:529

Amylotheca triflora, 5:182n49

Anacardium occidentale, 1:469n1

Anacardus, Ceylonese, 1:470

Anack, 4:12

Anaha Ewan, 2:449

Anahan, 1:348, 3:108

Anahul, 3:13

Anale, 5:495

Anamirta cocculus, 1:474n10,
  3:96n13, 3:141n8, 3:150n12,
  3:171n5, 3:527n1, 3:609n2,
  4:521n9

Ananas, 1:311, 1:313, 1:314, 1:317,
  4:151, 4:152, 4:153–4:156
  comosus, 1:341n39, 2:265n4,
    2:585n19, 4:151n1
  silvestris, 3:435
  wild, 4:224, 4:225

Anappur, 4:300
  ayer, 4:283

Anapul, 4:300

Anassa domestica, 4:151, 4:152,
  4:153–4:156

Anasser, 5:487

Ancou, 4:54

Andawas Tree, 2:129

Andjong atap, 5:72

Andjuri, 2:499

Andon
  watu, 3:315

Andong, 3:314

Andropagon
  aciculatum, 4:584n8
  amboinicus, 4:86n4
  citratus, 4:138n27

Andropogon
  aciculatus, 5:129*n3*
  amboinicus, 1:277*n17*, 5:215*n18*,
    5:234*n12*
  citratus, 5:29*n4*, 5:35*n2*, 5:38*n10*

Androsaceus crinis equi, 5:557*n2*

Andudu, 1:283

Aneilema
  nudiflorum, 5:43*n2*, 5:67*n7*
  vitiense, 5:45*n19*

Anemone, 4:246, 4:247–4:248

Angajin, 1:455

Angaro, 2:61

Angcoa, 4:373, 4:379

Angel plant, 5:171

Angelika, 1:349

Angiopteris amboinensis, 5:125*n1*,
  5:297*n2*, 5:303*n18*

Angraecum, 5:178, 5:179–5:185
  album, 5:197
    majus, 5:186, 5:187, 5:188,
      5:189
    minus, 5:189
  angustifolia, 5:199
  angustis crumenis, 5:201, 5:203
  caninum, 5:200, 5:201, 5:202,
    5:203, 5:204, 5:205–5:206
  decimum, 5:199
  furvum, 5:197
  lanuginosum, 5:206
  nervosum, 5:201, 5:203
  nonum, 5:197
  nudum, 5:204, 5:207, 5:208,
    5:209
  octavum, 5:197, 5:198, 5:199
  pungens, 5:203
  purpureum, 5:204, 5:207, 5:208,
    5:209
    silvestre, 5:209
  quintum, 5:193
  rubrum, 5:188, 5:191, 5:197
  saxatile, 5:203
  scriptum, 5:179, 5:215
  sediforme, 5:205
  septimum, 5:196
  sextum, 5:194, 5:195–5:196
  taeniosum, 5:206

  terrestre, 5:219
    alterum, 5:217
    primum, 5:213, 5:214, 5:215
      album, 5:215
      purpureum, 5:215
    tertium, 5:197
    uniflorum, 5:205

Angrec
  angin, 5:205
  cassian, 5:189
  cassomba, 5:209
  casturi, 5:195
  colan, 5:187
  gajang, 5:205
  jambu, 5:205, 5:209
  mera, *5:188*, 5:191
  puti
    besaar, 5:187
    kitsjil, 5:189
  red, *5:188*, 5:189
  tana, 4:245, 5:215, 5:217
  thonged, 5:206
  tsjampacca, 5:196

Angrecum scriptum, 4:563

Angrek, **5**:553
  andjing, 5:200, 5:201, 5:202,
    5:203, 5:204, 5:205–5:206
  calappa, 5:181
  cassumba, 5:207
  dog, 5:200, 5:201, 5:202, 5:203,
    5:204, 5:205–5:206
  ground, 5:209, 5:213, *5:214*,
    5:215, 5:217
  jambu, 5:207
  kitsjil, *5:224*, 5:227–5:228
    glap, 5:197
  kringsing, 5:181
    kitsjil, 5:193
  lemon kitsjil, 5:197
  lida, 5:181
  peasant, *5:208*
  primum, 5:217
  purple, *5:204*, 5:207, *5:208*,
    5:209
  pursed, *5:200*
  red, *5:188*, 5:191
  rock, 5:203, *5:204*
  single-flower, 5:205*n19*
  stinging, 5:203

tana, 5:223
  alea, 5:223
  itam, 5:223
  white, double, *5:186*, *5:187*,
    5:188, *5:189*
  yellow, *5:194*, *5:195–5:196*

Angrung, 3:399

Angsara javana, 3:669

Angune, 4:432

Anguri, 3:225

Anguria, 4:432
  indica, *4:430*, *4:431–4:434*

Anil, 4:146

Anipan dureyn, 5:266

Anise, 4:265*n10*, 4:393*n2*, 5:80

Anise Tree, Moluccan, 2:195,
  2:196, 2:197, 2:198

Anisum Moluccum, 2:195, 2:196,
  2:197, 2:198

Annanas, 4:153

Annas boaya, 4:225

Annil, 4:143, 4:145

Annipa waccan, 4:77, 5:271

Annona, 1:395, 3:107
  muricata, 1:340*n35*, 1:343*n1*,
    1:397*n12*, 3:244*n16*
  reticulata, 1:340*n35*, 1:348*n8*,
    1:395*n1*, 2:84*n13*, 3:107*n9*,
    3:569*n5*
  squamosa, 1:399*n1*, 1:404*n2*,
    4:68*n10*

Anoectochilus reinwardtii, 1:102,
  5:175*n1*

Anona, 1:395, *1:396*, 1:397–1:398
  maram, 1:398
  tuberosa, 1:398, 1:399

Anonis, 5:85

Anruda, 5:95

Ansjeli, 1:349, 1:353

Antaris toxicaria, 2:399*n15*

Anteuphorbium, 3:329, 3:331

Anthemis, 4:472

Anthistiria gigantea, 5:293*n3*

Anthocephalus macrophyllus,
  1:277*n19*, 2:324*n25*, 2:467*n1*,
  5:240*n5*

Anthophillis, 2:548

Anthophyllum saxeum, 5:439

Anthriscus, 4:211
  cerefolium, 2:71*n2*

Antiaris toxicaria, 1:56, 2:402*n21*,
  3:16*n6*, 4:134*n7*, 4:150*n14*,
  4:434*n21*, 5:302*n16*

Anticholerica, 3:273, *3:274*, 3:275–
  3:277

Antidesma
  bunius, 3:145*n1*, 3:145*n4*
  stipulare, 2:575*n1*, 3:147*n5*

Antidotum, 4:445

Antimon, 4:435
  ticos, *4:536*, *4:537–4:538*

Antipathes, 5:364
  dichotoma, 5:379*n1*
  ericoides, 1:383*n18*

Antsjac, 3:27

Anty, 5:119

Anxana Tree, 2:315

Ape-druiven, 4:519

Apepe, 3:491
  daun
    bezaar, 3:491, *3:492*, *3:493*,
      3:494–3:495
    kitsjil, 3:494

Api api, 5:181

Apium, 4:217, 4:217*n4*
  fruticans, 3:309
  graveolens, 3:309*n3*
  palustre, 4:219

Apocyna, 2:280, 2:454

Apocynaceae, 3:497*n1*, 3:497*n3*,
  3:523*n6*

Appacau, 5:481

Appau äu, 4:297

Apple, 3:91, 5:565, 5:566
  Assyrian, 4:6*n24*
  balsam, 4:447*n3*
  bitter, 4:426*n13*

bottle, 4:426*n9*
  of love, 4:455, *4:456*, 4:457
Appya, 3:134
Aquifolium
  indicum, 5:311, *5:312*, 5:313–
    5:314
  parvifolium, 5:313
Aquila Brava, 2:60
Aquilaria
  agallocha, 2:47*n1*, 2:84*n14*,
    2:87*n17*, 2:88*n33*, 3:501*n18*
  malaccensis, 2:47*n1*, 2:55*n14*,
    2:87*n18*, 2:88*n33*, 2:88*n38*,
    2:314*n7*, 2:316*n17*, 2:413*n25*,
    2:559*n6*, 3:270*n7*, 3:332*n16*,
    3:523*n5*, 3:559*n16*, 4:63*n37*
Aquilegia
  vulgaris, 5:213*n7*
Arabiae felicis, 3:567
Arabicum, 4:159
Arachis hypogaea, 4:473*n1*, 5:182*n50*
Arais el nil, 5:327
Arasavaali, 5:529
Araticu pone, 3:244
Arbor
  acida
    major, 2:509, *2:510*, 2:511
  alba, 2:543, 3:61, 4:321
    major, 2:103, 2:104, 2:105–
      2:107
    minor, 2:108, 2:109–2:110
  aluminosa, 3:57, *3:58*
  Brasiliensis, 2:227
  calappoides sinensis, 1:319, 1:320
  capraria, 2:528
  carbonaria, 3:401
  circinnorum, 2:570
  clypeorum, 3:119, 3:266
    alba, 3:87, *3:88*, *3:89*, 3:90
    rubra, 3:90
  coeli, *3:148*, 3:149–3:150
  conciliorum, 3:26, 3:27, *3:28*–
    3:29, 3:30–3:31
  exaecans, 3:167, 3:332, 3:411,
    3:501
  excaecans, 2:62, 2:363, 2:639
  exocecans, 2:542, 2:584

facum, 2:424
  major, 2:549, 2:550, 2:551,
    2:552
  minor, 2:553, 2:554
farinae ferax, 1:298
farinaria, 1:299
finium regundorum, 3:314
frigoris, 2:513
glutinosa, 3:47, 3:154
lacaria, 1:327
lactaria, 2:279, 2:365, 2:370,
  2:385, 2:386
Magellanica aromatica, 2:92
mangifera, 1:329
melonifera, 1:419
nigra, 2:420, 2:421–2:422
  latifolia, 2:425
  maculosa, 2:424–2:425
  parvifolia, 2:422, 2:424
noctis, 2:561
nuda, 2:574, 2:575, 3:147
nussalavica, 5:490, 5:493
ovigera, 3:121, 3:122, 3:123
palorum
  latifolia, 2:593
  nigra, 2:593, 2:595
  parvifolia, 2:593
pampajo, 1:419
pinguis, 1:461, 2:376, 2:377–
  2:378, 2:443, 3:134, 4:182
polyrrhizos, 3:5
pudica, 4:271
radicum, 3:5
radulifera, *3:138*, 3:139
raiz, 3:5, 3:7, *3:8*, 3:436
rediviva, 1:439, 3:67
regis, 2:389, 2:598
rubra, 1:369, 2:501–2:502, 2:543,
  2:544, *2:545*, 2:546, 2:549,
  3:119, 3:141, 3:141*n2*, 4:183
  angustifolia, 2:546
  prima, 2:543
  saxatilis, 2:547
  secunda, 2:547–2:548
salamandra, 3:147
sebi, 5:476
sinium regundorum, 2:117
spiculorum, *3:70*, 3:71–*3:72*
  aeruginea, 3:71
  augustifolia, 3:71
  larifolia, 3:71

suaviter olens, 2:245

toxicaria, 2:375, 2:396, 2:397–
    2:404, 5:302

tristis, 3:559

vernicis, 2:391, 2:392, 2:393–
    2:395

versicolor, 2:634, 2:635

vesperilionum, 5:499

violaria, 2:309

Arbore de Raiz, 3:5, 3:8, 3:436

Arbore de sombreiros, 1:249

Arbusculas marinas, 5:357–5:360

Arbutus

    menziesii, 3:425*n3*

    molucca, 3:425

    unedo, 3:425*n3*

Arcangelisia flava, 3:533*n1*,
    4:478*n16*

Arealu, 3:30, 3:31

Areca, 1:223–1:230, 3:119*n3*,
    4:309*n3*, 4:362*n16*, 5:389*n41*,
    5:396n8

    catechu, 1:223*n1*, 1:339*n31*,
        2:258*n19*, 2:300*n3*, 2:555*n2*,
        3:27*n3*, 4:125*n14*, 4:173*n5*,
        4:409*n5*, 5:126*n11*, 5:217*n2*,
        5:222*n13*

    glandiformis, 1:241*n6*

    nut, 5:15*n14*, 5:51*n16*

    pinnata, 4:287*n16*

    vestiaria, 1:244*n29*

Aréca goli, 3:24

Arenga

    pinnata, 1:56, 1:212*n19*, 1:241*n8*,
        1:273*n8*, 1:340*n33*, 2:133*n6*,
        2:181*n15*, 2:204*n9*, 2:258*n21*,
        2:267*n11*, 2:623*n6*, 3:258*n13*,
        3:258*n15*, 3:409*n7*, 3:409*n8*,
        3:639*n12*, 4:78*n5*, 4:123*n2*,
        4:182*n44*, 4:203*n3*, 4:297*n5*,
        4:449*n8*, 4:565*n2*, 4:572*n4*,
        5:34*n12*, 5:153*n7*, 5:193*n8*,
        5:215*n20*, 5:259*n6*, 5:341*n6*,
        5:459*n3*

    saccarifera, 4:25*n3*

    saccharifera, 5:193*n8*

Argawân, 4:197

Aria-decca, 1:225

Ariadne's crown, 4:539, *4:540*,
    4:541

Arian, 4:561

Arien

    essong, 5:345

    malona, 5:345

Arisarum

    amboinicum, 4:293, 4:295–4:296

    esculentum, 4:277, 4:297, *4:298*

Aristolochia

    rumphii, 4:557*n1*

    tagala, 4:555*n1*

    tertia, 4:507

Armatoira, 1:319

    das Igresias, 1:320

Arom esculentum, 5:335

Aros patsjari, 4:109

Arreboleras, 4:192

Arroz, 4:110

Artabotrys suaveolens, 3:503*n1*,
    3:605*n5*

Artemisia, 3:294, 5:519*n4*

    abrotanum, 5:429*n6*

    argyi, 4:203*n7*

    latifolia, 4:198, 4:203–4:204

    vulgaris, 4:201*n16*, 4:203*n1*

Articu, 1:397

Artocarpus, 1:352*n7*

    altilis, 1:355*n1*, 2:350*n9*,
        2:391*n8*, 2:446*n7*, 3:175*n2*

    champeden, 1:347*n1*, 1:350*n1*,
        3:4*n6*, 3:451*n2*, 5:237*n2*,
        5:237*n4*, 5:353*n4*, 5:473*n6*

    communis, 1:350*n3*, 1:417*n5*,
        2:138*n10*, 2:467*n6*, 2:503*n5*,
        3:41*n5*, 3:118*n18*, 3:382*n11*,
        4:29*n7*, 4:473*n6*, 5:451*n6*

    pungens, 1:357*n1*

    dasyphylla, 2:447*n3*

    elastica, 3:463*n7*

    elasticus, 1:357*n1*

    fretissi, 1:359*n5*, 2:447*n3*

    heterophyllus, 3:528*n11*

    hirsutus, 1:349*n13*

integer, 1:343*n1*, 1:347*n2*,
    1:350*n1*, 2:447*n1*, 2:575*n8*,
    3:107*n8*
integra, 1:317*n14*, 1:339*n25*,
    5:237*n4*, 5:366*n20*
integrifolia, 4:4*n9*
    reticulata, 1:359*n5*
Aru palakka, 3:127
Arum, 4:288*n24*
    aegyptiacum, 3:81, 3:117
    aegypticum, 4:285, 4:286, 4:287–
        4:290, 4:288, 4:291, 4:292,
        4:293
    aegyptium, 5:321
    aquaticum, 4:281, 4:283, 4:284
    domesticum, 4:278
    dragon, 4:282
    hortense, 5:301
    indicum
        sativum, 4:276, 4:277–4:279
        sylvestre, 4:281
    latifolium, 4:281
    maculatum, 4:290*n41*
    medium, 4:281
    sativum, 4:277
    silvestre, 4:291
    thieves, 4:282
    vulgare, 4:281
    water, 4:283, 4:284
    wild, 4:280, 4:281–4:282
Arundarbor, 3:180, 3:181–3:185
    aspera, 3:181, 3:191, 3:195
    cratium, 3:187–3:188
    farcta, 3:211, 3:211*n2*, 3:212,
        3:213
    fera, 3:181, 3:203, 3:204, 3:205–
        3:207
    maxima, 3:181, 3:191, 3:195,
        3:197
    nigra, 3:206
    spiculorum, 3:181, 3:189–3:190
    spinosa, 3:181, 3:199, 3:200–
        3:201, 3:202
    tenuis, 3:181, 3:184
    vasaria, 3:181, 3:191, 3:192,
        3:193*n24*
Arundastrum, 3:214, 3:215–3:216
Arundinella, 5:175
    aquatica, 5:44, 5:45

latifolia, 5:175
    minor, 5:40, 5:43–5:45
Arundo
    arenaria, 5:33*n6*
    farcta indica, 3:642
    indica
        arborea, 3:180, 3:181–3:185,
            3:193, 3:642
        farcta, 3:207
            nastos, 3:637, 3:638, 3:639–
                3:642
        japanica, 3:207
        saccarifera, 4:92, 4:93–4:97
        toxica, 3:211, 3:212, 3:213
Arupa, 3:143
Arupa Tree, 2:524, 2:525
Arvore da raspa, 3:406
Aryan harulu, 5:345
Aryophyllaster ruber, 3:162
Asarath, 4:125
Asarum europaeum, 4:523*n4*
Asclepadaceae, 3:491*n2*, 3:495,
    3:495*n10*, 3:497*n3*
Asclepias
    curassavica, 5:101*n2*, 5:495*n3*
    syriaca, 5:101*n2*
Aspidium repandum, 3:317*n10*,
    5:133*n5*
Asplenium, 2:222*n28*, 5:129*n14*,
    5:153*n1*, 5:157*n1*
    adiantum-nigrum, 5:140*n14*
    nidus, 5:153*n2*, 5:153*n4*
    scolopendrium, 5:364*n5*
    trichomanes, 5:145*n2*
Assa
    ihe, 3:605
        metten, 3:605
    wali, 3:103
Assaïe, 4:491
    tuni, 4:494
Assam-Java Tree, 2:133
Assar, 5:315
Asseut, 5:257
Assi Assi, 2:623

Assotelina, 5:49, *5:50*, 5:51, *5:52*, 5:53

Assuseng batu, 5:445

Assuseng Tree, 2:279

Assyrian Apple Tree, 2:140

Assyrian Plum, 3:49*n12*

Astonia, 2:542*n11*

Astonishment plant, 5:73, *5:74*, 5:75

Astronia papetaria, 3:421*n1*

Ata-maram, 1:401

Athyrium
  esculentum, 3:100*n4*, 4:471*n16*, 5:127*n1*
  filix femina, 4:270*n30*

Atis, 1:399

Atractylis, 4:135

Atrebatis, 5:430*n12*

Atsjar, 4:186

Attehu, 1:420

Attihu selu, 5:150

Atty alu, 3:5, 3:8

Atun
  laut, 1:465, 2:588
  tree, *1:458*, 1:459–*1:460*
    white, 1:460*n9*
  utan, 2:587

Atunua racemosa, 1:459*n1*, 5:340*n10*

Atunus, 1:460
  litorea, 2:586, 2:587–2:588, 2:602

Aupalulan, 3:359, 3:363
  mahina, 3:365

Aupalulang, 5:259
  puti, 5:259

Auparulan, 4:522

Aurantium, 2:163, *2:164*, 2:165, 2:169
  acidum, 2:163, *2:164*, 2:165
  pumilum Madurense, 2:171
  Sinense, 2:166, 2:167–*2:168*
  verrucosum, 2:169, *2:170*, 2:171

Aurgading, 3:203

Auricularia auricula-judae, 5:243*n16*, 5:346*n9*

Auris
  canina, 5:80
    femina, 5:49, *5:50*, 5:51, *5:52*, 5:53
    mas, 5:49, *5:50*, 5:51, *5:52*, 5:53
  elephantis, 5:437

Aussi abbal, 3:507, 3:508

Aussi Pipis Tree, 2:155

Authau, 4:409

Avanacu, 3:339, 3:340

Avbibi Tree, 2:205

Avellana Indica, 1:225

Avellanae purgatrices, 3:347

Averrhoa
  bilimbi, 1:365*n1*, 2:125*n23*, 2:207*n5*, 2:309*n4*, 2:577*n6*, 3:83*n3*, 3:151*n2*, 3:431*n4*, 3:623*n3*, 3:631*n2*, 5:529*n3*, 5:531*n2*, 5:569*n5*
  carambola, 1:289*n3*, 1:333*n3*, 1:361*n1*, 2:513*n6*, 5:180*n22*

Averrunca spectrorum, 5:60

Avicennia officinalis, 1:361*n1*, 2:623*n1*, 2:635*n5*

Awackal, 3:103

Awar Awar, 3:45

Axi, 4:189

Axis, 4:125

Ay Alla, 2:635

Ay Goa, 2:87

Ay Hua Mette, 2:279, 2:280

Ay lala, 5:497

Ay Lapia, 3:143

Ay Macuninu Tree, 2:207

Ay Mousiën, 2:567

Ay nou ollo, 1:420

Ay puti, 1:495, 1:497

Ay somin, 5:493

Ay-alat Aussi, 2:203

Ay-Hoa-Marua, 2:300

Ayam bakuku, *3:620*, 3:621–3:622
Aycamenil, 2:67
Aycau, 3:119, 3:141, 3:161, 3:162
  Boppo, 2:547
  boppo, 3:141
  Bugulawan, 2:547
  Hatulale, 2:547
  LauIla, 2:548
  laun maun, 3:161
  tuha, 3:141
  Uhel, 2:543, 2:546
Aycau laun ela, 1:379
Aycora Tree, 2:89
Aycou, 1:379, 1:381
Aydjuan, 4:219, 4:220
Ayer pera, 4:291
Ayewan, 2:489
Ayhau Madelle, 2:245
Ayhohitt, 4:101
Ayhoït, *5:40*, 5:41
Ayhola Lamahu, 3:69
Ayhua naru, 5:158
Ayhua-Laha, 2:575
Ayhuäa, 3:329
  besaer, 4:228
  kitsjil, 4:228
  mahina, 4:228
Ayile, 5:140
Aylanto, 3:150
Aylanu
  hosso, 3:121
  sclu, 3:72
  sunu, 3:81
Aylaru, 3:84, 3:567
  pohon, 3:85, 3:567
Aylassalinu Tree, 2:207
Aylatu, 2:389, 4:521
Aylau sala, 4:515
Aylauessi, 2:549
Aylaun
  ahussen, 5:65, 5:67
  aitu, 4:255
  capepuli, 4:523
  hitu, 4:258

  laut, 3:423
  Maccur, 2:473
  mahay, 4:268
  mamulin, 5:60
  manu, 4:237
  marva, 3:389
  nya, 3:581
  peu peu, 5:65
  pia sanaf, 3:427
  seron, 5:575
  tahina, 4:255, 4:258
  tahinan, 5:108
  tahur, 5:259
  ua massarii, 5:65
  Umali, 2:473
  waccan, 4:529
Aylausiën, 5:485
Aylautassi Laun Maun, 2:515
Aylilin utan, 5:63
Ayloaha, *5:56*, 5:57–5:58
  utan, *5:56*, 5:59–5:61
Aylohoy, 3:301
Aylohun
  payamatta, 4:534
  ubi, 4:507
  uyn, 4:529
Aylou
  aha, 5:95
  tuban, 3:257
Aylouha puti, 5:99
Ayloun
  ahussen toani, 5:71
  assa tree, 5:505
  mamuli, 5:61
  marua, 3:371
  niwel, 3:301, 3:305
  puti, 5:111
  sehu, 5:65
Aylounya, 5:63
  pohon, 5:63
Aylous Hosso, 3:127
Aymaan Abbar, 2:508
Aymacca lulan, 3:359
Aymahu, 1:424
  abba, 1:425
Aymalaha, 2:496, 3:90

Aymalu, 3:27

Aymalua, 3:391

Aymanu, 3:60

Aymette, 3:109

Aymette Paleku, 2:420, 2:421

Aymetten, 2:421
  Daun Kitsjil, 2:422, 2:423
  Lou Yla, 2:425
  Paleku, 2:424

Aymiri, 2:389

Aymiten, 2:412, 2:419, 2:595

Aynahu, 3:427

Ayncca, 3:351

Aynier, 1:420

Aynitu, 3:407

Aynywer, 1:420

Ayocor, 3:345

Aypacca, 3:47
  java, 3:27

Aypaloulan, 2:552

Aypolat, 3:176

Aysalo, 2:521

Aysamara, 2:570

Aysana, 5:339

Aysaru, 2:67

Aysielo, 3:135

Aysuli abbal, 3:291

Aytaël LayIla, 2:555

Aytassi, 2:515

Aytette
  laun Maun, 3:289
  obtte, 3:286
  tuban, 3:257

Aytiba, 3:674

Aytimon, 3:173

Aytimul Tree, 2:518, 2:519

Aytohi, 3:167

Aytssana, 5:339

Aytua, 2:631

Aytuban, 3:257, 3:259

Aytuy, 3:167

Ayucca, 3:355

Ayulan, 1:451

Ayulet, 2:517

Ayulit, 1:451

Ayune, 3:147

Aywäa umali, 3:553

Aywahäu, 4:83

Aywara, 3:671

Aywayl, 3:422, 3:423

Aywel, 4:358

*Azadirachta
  excelsa*, 2:643n1

B

Babalu bugis, 5:555

Babesarang, 5:478, 5:479

Babon, 3:154

Babora, 4:429

Babuta, 2:363

Bacan, 1:39

Baccaurea nanihua, 2:437n27,
  5:468n1

Backla jemanija, 4:159

Bacompol, 3:584, 3:585–3:586

Bacong, 5:301

Bacovo, 4:6, 4:9n6, 4:10, 4:11

Badjang badjang, 5:23, 5:24

Badlingan, 4:375

Bado, 1:429

Badong, 2:380

Baggea Leaf
  beach, 3:455, 3:456, 3:457
  true, 3:453

Baham louhur, 4:165

Bahingin, 3:8

Bahu lawan abbal, 2:21

Bahuta, 4:312

Baingui, 3:24

Bajan
  besaar, 4:165
  utan, 4:161

Bajang
  baduri, 4:161
  bonga, 4:168
  cabeçette, 4:168
  cassomba, 4:168
  Chinese, 4:157
  eckor kutsjing, 4:168
  mera, 4:157
  pohon, 4:165
Bajanta, 4:161
Bakar, 1:352
Bakewe, 2:62
Bako, 2:617
Baku teker, 5:11, *5:12*, 5:13
Balabi, 3:154
Baladong lengis, 3:331
Balan
  djingan, 2:497
  tronga, 2:293
Balaustium, 2:137
Balintette, 4:477
Balis Buckol, 2:174
Balla
  balla, 5:335
  modagam, 3:382
Ballota nigra, 4:255*n3*, 5:73*n6*
Balsam
  apple, 4:447*n3*
  pear, 4:447*n1*
  plant, 5:291
Balsamina
  altera, 4:*194,* 4:195–4:197
  amygdaloidem, 4:197
  femina, 4:*194,* 4:195–4:197,
    4:197
    indica, 4:196
  foemina, 3:251
  mas, 4:447, 4:449
  persicifoliam, 4:197
Balulangh, *5:16,* 5:17–5:18
Bama, 5:353
Bambagia, 3:234
Bambalu, 3:476
Bamban, 3:215
Bambasinen, 3:234

Bambax, 3:234
Bamboesen, 3:181
Bamboo, 1:51, 1:136, 3:72, 3:181,
    3:181*n1*, 3:181*n6*, 3:181*n7*,
    3:182, 3:188*n8*, 3:189*n8*, 3:193,
    3:195*n1*, 3:202, 3:216, 3:331*n7*,
    3:507*n8*, 3:657, 3:669, 4:66
Bambusa
  atra, 2:273*n6*, 3:182*n12*, 3:184
  bambos, 3:199
  blemeana, 1:51
  blumeana, 5:21*n6*
  excelsa, 3:195*n5*
  spinosa, 3:199*n1*
  vulgaris, 2:414*n31*, 3:191*n2*,
    3:193*n24*, 3:203*n1*, 3:272*n30*,
    3:642*n27*, 4:95*n5*
  striata, 3:203*n3*
Bamia, 3:244
  alpini, 3:244
Bana, 4:494
Banana, 1:142, 1:204*n33*, 1:395*n3*,
    1:417*n12*, 1:507*n19*, 2:635, 3:9,
    3:107*n6*, 3:154*n8*, 3:533*n2*,
    4:3–4:7, 4:3*n1*, 4:5*n16*, 4:277*n3*,
    4:322*n27*, 5:262*n9*, 5:531*n1*–5:3
  forest, 4:15*n5*
  king, 4:11*n21*
  red, 4:11*n22*
Bancal, 2:563, 2:564, 3:52
  parampuan, 2:561
Bancalus
  femina, 2:561
  latifolia, 2:561
  mas, 2:563, 2:564
  parvifolia, 2:563, 2:564
Banckal Lacki Lacki, 2:563, *2:564*
Bancudas
  angustifolia, *3:50,* 3:51–3:52
  latifolia, 3:53, *3:54,* 3:55
Bancudi tree
  broad-leaved, 3:53, *3:54,* 3:55
  narrow-leaved, *3:50,* 3:51–3:52
Bancudu, 2:79, 2:81, 3:57
  daun Bezaar, 3:53, *3:54,* 3:55
  root, 3:271
Banda, 1:39, 1:41, 1:43, 1:59

Bandira, 3:27

Bandong, 4:140

Bangcoan, 4:373

Banghian Tree, 2:189

Bangian lacki lacki, 5:569

Bangle, 3:294, 5:471, 5:511, 5:555,
    5:559*n4*
    Malacca, 4:77
    malacca, 5:267, 5:271, 5:273
    shrub, *4:44*, 4:47–4:48, 4:63,
        4:67, 4:84, 4:138*n31*

Bangon bangon, 4:258

Bangue, 4:125, 4:127

Bangung, 3:100

Bangyan lacki lacki, 5:539

Bangyan Tree, 2:185

Banka, 1:34

Bantam, 1:37, 1:42, 1:49

Bantiala, 5:231

Bantohoae, 4:181

Banyan, 3:30*n14*, 5:182*n48*,
    5:208n9, 5:240n6

Bapasay, 3:471

Bappa silan, 3:571

Bara mareka, 4:387

Barba
    laut, 5:457
    marina, 5:457
    saturni, 5:167

Bariga, 4:146

Barleria prionitis, 5:511*n1*

Barokokalos, 4:179

Barringtonia
    asiatica, 3:95*n1*
    racemosa, 3:97*n1*

Baru, 1:283, Tali2:335
    creeping, Tali2:335
    Laut, 2:340
    Partey, 2:340
    tschina, 4:203

Barudan Aleu, 3:139

Barus Tree, 2:127

Basilicum, 4:205, 4:206, 4:207
    agreste, *4:207*, 4:209–4:210,
        4:210
    album, 5:293
    Arabian, 4:207
    caryophyllum, 4:209
    hortense, 4:205
    sulassi, 5:293
    vulgare, 5:293

Baso, 5:129

Basta
    laut, 5:451
    marina, *5:450*, 5:451

Bastaart Ebben-hout, 2:344

Bat tree, *5:498*, 5:499

Bat's Greens, 4:451

Bata, 1:329

Batang, 1:338

Batari, 3:501

Batatta, *4:368*, 4:369–4:372
    mammosa, 4:373, *4:374*, 4:375,
        4:379
    sinensis, 4:477

Batavische hagedoorn, 5:539

Bate, 1:333

Bateca, 4:429

Batel, 1:333

Bathyriza, 5:86

Bati bati, 5:73

Bats' leaf, 4:454

Batsjam batsjam, 4:138, 4:528

Batsjan, 1:333

Batsjang kiang, 5:309

Batsjar, 3:251, 4:196
    petok, 3:251

Battari, 4:105

Battattas, 5:235

Batteca, 4:432, 4:434

Battich hindi, 4:431

Batticho, 4:435

Batu
    barudan, 5:444
    malacca, 1:333

swangi, 5:423
ticcar, 5:446
ubi, 5:446
Bau mautu, 5:371
Bauhinia, 3:469
   lingua, 1:277*n21*, 2:427*n3*,
    3:467*n3*, 4:475*n4*
Bauraka, 3:134
Bauwa
   appar, 5:307
   furu, 5:307
Bauzoz-Indo, 1:196
Bawang
   puti, 4:312
   utan, 5:305, 5:307
Baxilicum aquatile minus, 5:291
Bay nan tyade, 1:365
Baya, 3:53, 3:55
Bayang, 4:159*n11*
Bayen wood, 2:467
Beach Atun Tree, 2:586, 2:587–
   2:588
Beach Casuaris Tree, 2:568, 2:569–
   2:571, 2:572, 2:573
Beach Granate Tree, 2:580, 2:581–
   2:582, 2:583, 2:584–2:585
Beach jasmine shrub, 3:617, 3:618,
   3:619
Beach Kellor tree, 3:135
Beach Pepper, false, 3:417*n7*
Beach pod, 3:481, 3:482, 3:483
Beach shrub, bitter, 3:595
Beach Tree, 2:323*n14*
Beach Virgin's Herb, 3:481
Beach-Clove tree, wild, 3:368,
   3:369
Bead tree
   broad-leaved, 3:85, 3:86
   small-leaved, 3:82, 3:83–3:84
Bean, 4:391–4:392, 5:39*n20*
   Chinese, 4:383, 4:384, 4:385,
    4:401
   earth, 4:474
   Egyptian, 4:288*n24*

   French, 4:379*n3*
   golden, 4:405*n1*
   green, 4:405*n1*, 4:405*n3*
   Indian, 4:359, 4:399
   Indian three-leaved, 4:392
   kidney, 4:379*n3*, 4:383*n2*
   little black, 4:409*n3*
   little red, 4:401*n5*
   mountain, 5:280, 5:281–5:282
   mouse, 4:406*n11*
   Roman, 4:379*n3*
   scripted, 4:399*n1*
   soy, 4:409*n1*
   sticky, 5:289
   van der Lyns, 4:385
   velvet, 4:395*n1*
   white, 4:401*n2*
   yam, 4:379*n1*
Beech Tree, 2:3
Beechwood, 2:63
Beet-klim, 4:459
Beetla codi, 4:323
Begonia, 4:138*n41*, 4:229I*n1*,
   4:561*n9*
   tuberosa, 4:527*n1*, 4:527*n5*,
    5:175*n5*
Beid el ossar, 5:514
Bejettan, 1:424
Bel ericu, 5:514
Bela schora, 4:429
Belawa, 2:393
Belilla, 3:372
Belimbing buloh, 2:309*n4*
Belle de Nuit, 4:191*n1*
Bellis, 4:249, 5:83*n3*
Bellonius, 4:431
Belous, 1:506
Beludru, 4:437
Beluru, 3:227
Bem munja, 1:496
Bencaro, 3:78
Bencuda Lakki Lakki, 3:51
Bengal panick, 4:116

Benincasa hispida, 1:418*n14*,
   4:421*n1*, 4:435*n5*

Benjoin, 4:63

Bera pulut, 4:113

Beram, 4:136

Berang, 1:196

Beras, 4:110

Berbuz, 4:159

Berry tree, *3:14*, 3:15–3:16, *3:17*,
   3:18
   Brasilian, 3:421
   wild, 3:19, *3:20*, 3:21

Berry-bearing tree, *3:352*, 3:353
   East Indian, 3:372, 3:426

Beseeching plant, *5:208*, *5:210*,
   5:211–*5:212*

Bessa magumi, 4:556, 4:559

Betele, 4:323

Betonica fruticescens, 4:168

Betre, 4:320

Betris, 4:189

Betzy, 4:354

Beursjes angrec, 5:189

Bia
   baru, 5:399
   bia, 5:335
   garu, 5:416

Biarera, 4:110

Biarum, 5:327

Biawas, 1:379

Bibio maman, 3:311

Bidac bankong, 5:65

Bidara Tree, 2:178

Bidens chinensis, 1:65, 5:71*n1*,
   5:279*n4*

Bido
   marau, 4:320, 4:326
   masoffo, 4:326

Bidsjam
   hitem, 4:122
   puteh, 4:122

Bierblad, 4:449

Bifi-mafalla, 2:389

Big Parsley tree, 3:311*n4*

Bihaun, 4:354

Bilac carbou, *1:508*, 1:509

Bilacus
   bilac, 1:506
   taurinus, 1:509

Bilak tree, *1:504*, 1:505–1:507
   carbau, 1:505
   pissang, 1:505
   telloor, 1:505

Bilalangh Tree, 2:129

Bilimba, 1:362

Bilontas, 4:265

Binalo api, 5:173

Binan, 1:226

Bindatöe, 4:389

Bindaus, 2:374

Bindweed, 3:467, 3:564*n16*,
   4:192*n7*, 4:467, 4:545
   blue, 4:483
   greater, 4:475, *4:476*, 4:477–
     4:478
   lesser, 4:479, *4:480*, 4:481
   maderaspatanian, 4:478
   marine, *4:484*, 4:485–4:486
   prickly, 4:335*n2*, 4:491, 4:503
   riverside, *4:484*, 4:487
   rough, 4:491*n1*
   smooth, 4:475, *4:476*, 4:477–
     4:478, 4:479, *4:480*, 4:481
   stinking, *4:488*, 4:489–4:490
   thorny, 4:497

Bingcawan, 5:159

Bintangor, 3:103
   Daun Kitsjil, 2:329
   Laut, 2:322
   laut, 1:387*n2*
   maritima, 2:321–2:326, 5:469,
     5:471*n20*
   montana, 2:327, *2:328*, 2:329,
     2:525
   mountain, 2:327, *2:328*, 2:329
   sea, 2:321–2:326
   silvestris, 2:577
   tenufolia, 2:329
   utan, 2:329, 3:143
   wild, 2:327, *2:328*, 2:329

Bintano, 3:113

Bintaro, 2:371

Bio bio, 5:251

Biophytum sensitivum, 3:266n8,
3:635n2, 4:267n2, 4:269n16,
5:79n2, 5:79n5, 5:129n5,
5:283n2

Bipula, 3:30

Bira, 4:110
ayer, 4:283, *4:284*
castela, 4:115
kitsjil, 4:295
negry, 4:278
pantsjuri, 4:282
utan, 4:281

Birala mado, 1:283

Birani, 3:34

Birch, 3:229, 3:230

Bird tree, 3:60

Bird's nest
edible, *5:336*, 5:341–5:344

Birindjene, 3:617

Birkas, 5:327

Birurong mera, *3:424, 3:425–3:426*

Bischofia javanica, 5:561n6

Bisel nil, 5:327

Bissi mattan, 4:523

Bissula, 1:310

Bitangor, 3:95

Bitbit, 5:135
besaar, 5:133

Bitter plant, 4:447, *4:448*, 4:449,
4:450, 4:451

Bitter sajor, *3:596, 3:597–3:598*

Bitule, 4:361

Biturong itam, *3:427, 3:428*, 3:429

Bixa orellana, 2:115n1, 3:314n4,
4:163n5, 5:207n3

Black Dammar Tree, 2:239, 2:240–
2:241, 2:242

Black Tree, 2:420, *2:421–2:422*
broad-leaved, 2:425
narrow-leaved, *2:420, 2:421*

small-leaved, 2:422, *2:423*
speckled, *2:420, 2:421*

Black Wood, 2:412

Blackberry shrub
Moluccan, *3:620, 3:621–3:622*

Blackthorn tree, 1:385n3

Bladder wort, 5:115, *5:116*, 5:117

Blechnum orientale, 5:133n1,
5:304n29

Blessed Thistle, 3:245n2, 3:309n2

Blimbing, 5:180n22
assam, 1:362
bessi, 1:362
bulu, 1:365
iron, 1:362n12
kris, 1:362
manis, 1:362
teres, 3:83

Blimbing tree, *1:360*, 1:361–1:364
long, 1:365, *1:366*, 1:367–1:368
sour, 1:365, *1:366*, 1:367–1:368
wild, *3:430, 3:431*

Blimbingum
silvestre, 3:430, 3:431
teres, 1:365

Blimbinos, 1:363

Blimbven, 1:363

Blinbinjum teres, 3:151

Blind Wood Tree, *2:358*, 2:359–
2:361, 2:362, 2:363, 2:370

Blinding shrub, *3:410*, 3:411–3:412

Blinging utan, *3:430*, 3:431

Blint-hout, 2:363

Blister producing root, 4:521n4

Blitum, 2:406, 4:157, *4:158*, 4:159–
4:160
agreste, 4:161, *4:162*, 4:163
album, 4:157
amboinicum, 4:157
fruticescens, 4:162, 4:165, 4:166,
4:167–4:168
maculosum, 4:157
Peruvianum, 4:159
repens, 4:159
rubrum, 4:157

spicatum, 4:160
spinosum, 4:161, *4:162*, 4:163
terrestre, 4:159
Block-Zedoar, 4:67
Blombing bulu, 3:623
Blood
  herb, *4:568*, 4:569
  sajor, *4:568*, 4:569
Blood-flower, 5:101*n2*
Bloodwort, mouse, 5:285
Bloody flux Tree, 2:27
Blue buttons, 4:249*n2*
Blue Clitoris Flower, *3:562*, 3:563–
  3:564
Blumea
  appendiculata, 5:107*n6*
  balsamifera, 1:408*n4*, 2:374*n10*,
    3:257*n9*, 5:107*n2*, 5:257*n2*
  chinensis, 4:263*n1*
Blycko, 4:421
Boa
  arum, 5:565
  Assuseng, 2:280
  ati, 1:399
  Balangan, 2:198
  Buton, 3:96
  coan, 1:438
  Cras, 2:271
  Djarong, 2:428
  Djudjaru, 2:300
  frangi, 1:469
  gore, 3:625
  Lawan, 2:7
  malacca, 5:465
  massy, 5:472, 5:473
  mulo, 5:573
  Nire, 2:582
  Parang, 2:442
  pausengi, 5:384, 5:385
  puttar, 5:525
  radja, 5:525
  rau, 1:435, 5:476, 5:493, 5:493*n2*
    utan, 1:438
  Rau-Utan, 2:378
  Sabon, 2:199
  sengi, 5:384, 5:385
  Sima, 3:60

Taator, 2:582
Tay Cambing, 2:575
tsjangurit, 1:227
tsjeremey, 5:529
Boano, 1:38
Bobata, 3:531
Bobbi Tree, 2:276
Bobe Fruit tree, *3:68*, 3:69
Bobo, 1:290
Bobora
  calbassa, 4:426
  dagoa, 4:423
Bobori, 4:62, 4:66
  hot, 4:63
Bobuto, 3:219
Bocke Sajor, 3:153
Boddam, 4:115
Bodelha, 5:346
Bodhi tree, 3:27*n7*
Bodjo Bodjol, 3:211
Boehmeria nivea, 2:405*n4*, 4:133*n1*
Boekewist, 5:251
Boën api, 3:490
Boerlagiodendron palmatum,
  3:349*n1*, 3:578*n4*
Boha-Hari Tree, 2:194
Boils leaf, 4:479
Boisua, 1:455
Bojama, 4:153
Bokyni coteli, 3:564
Bolehulan, 4:300
Bolematta, 3:108
Boletus
  moschocaryanus, 5:239
  primus
    infundibuli
      figura, 5:241
      forma, 5:240, 5:241–5:243
  quintus, 5:243
  saguarius, 5:239
  secundas arboreus, 5:241, 5:242
  terrestris, 5:242, 5:243
  tertius, 5:242

Bolintas, 4:263

Bololo Mahul, 2:473

Bombo terbang, 5:187

Bombori, 4:485

Bona allamoda, 4:399

Bonckus Cuman, 2:473

Bondacalo, 3:244

Bondor, 4:421, *4:422*, 4:423

Bonelau, 4:41

Bonga
  balitsjaja, 3:223
  biru, *3:562*, 3:563–3:564
  boki, 5:181
  calente, 3:563
  Cananga, 2:296
  castela, 4:247
  Eydja, 2:303
  fangan, 3:16
  gulong tsjutsju, *3:324*, 3:325–
    3:326
  haram tsjade, 4:247–4:248
  Hollanda, 4:466
  japon, 5:495
  jawa, 4:466
  knop, *4:246*, 4:249–4:250
  labe, 3:543
  Lacca, 2:639
  manilha, 3:327, *3:328*
  manoor, *3:556*, 3:557–3:561,
    3:560, 3:563, 4:521
    tudju lapis, 3:557
  manor, 5:433*n22*
  matta hari, 4:247, 4:248
  merak, 3:266
  nassi, 3:563
  pangil, 3:363
  pavaon, 3:266
  petang, 3:225
  pitsja, 3:365, *3:366*, 3:367
  puan, *3:362*, 3:363–3:364
  putri, 5:181
  raja, 3:217, 3:294, 4:510
    puti, 3:327, 3:559
    red, 3:220
    white, 3:217, 3:219
  siam, *5:570*, 5:571
  soussong, 3:327, *3:328*
  susanna, 4:245

tali, 4:466

tallan, 3:563

tanjong, 2:171, 2:287, *2:288*,
  2:289–2:290, 2:322, 2:394,
  2:410, 2:434, 2:435, 2:549,
  2:607, 2:625, 3:249
  Laut, 2:291

taratti, 5:321, 5:327

tay ayam, 3:560

tiga lapis, 5:221

tsjana, 3:266

tsjupo, 4:247

wacta, 3:223, 3:230
  cuning, 3:225, *3:226*, 3:227

wari, 3:327

wungan, 3:558

Bonga raja, 1:505

Bongac njingin, 3:327

Bongo bongo, 3:295

Bongoc Alas Tree, 2:123

Bonkos cussu, 3:297

Bontal, 4:455

Bonti, 4:497

Bonzo Chapete, 3:84

Boom-varen, 5:122

Boomwol, 3:234

Bootjungi, 5:384

Bootsja, 4:366

Boppa tsjeda, 5:304

Boppa tzeda, *3:380*, 3:381–3:383

Boppo, 5:133, 5:135

Boproôn, 3:527

Borage, 4:249, 4:249*n4*

Borago officinalis, 3:555*n4*

Borassus, *1:268*, 1:269
  femina, 1:261
  flabellifer, 1:204*n31*, 1:205*n35*,
    1:212*n21*, 1:256*n8*, 2:258*n20*,
    2:333*n5*, 3:645*n2*, 3:660*n12*,
    5:397*n11*, 5:425*n7*, 5:507*n4*,
    5:525*n2*
  mas, 1:261

Bori, *3:344*, 3:345–3:347
  bidji, 5:573

Boribori root, 4:60

Borium, 3:171

Boruro, 3:345

Bosluysen, 5:23, *5:24*

Botatto, 5:501

Botolino, 2:412, 2:415

Botor, 4:381, *4:382*

Botton, 4:115

Bou batati, 4:466

Boundary shrub, *3:312*, 3:313–
    3:314
    small-leaved, 3:315, *3:316*, 3:317

Bour Hanit, 2:311

Boutan, 3:220

Boxhoorn, 2:542

Boxwood, 3:514, 3:516, 3:613*n5*

Boy
    öa hoë, 3:266
    tsy, 4:499, 5:235, 5:235*n30*

Brahmin Beans, 3:564*n19*

Branbang utan, 5:309

Brappat, 1:389, 2:615, 2:617, 2:639,
    2:640, 2:641
    Gelang, 2:643
    Kitsjul, 2:627
    Tudong, 2:639

Bras
    djintam, 4:109
    pudac, 4:109
    pulut, 4:113

Brasil Wood, 2:76, 3:271, 3:272,
    3:272*n35*

Brassica
    juncea, 3:619*n12*, 4:235*n1*
    marina, 4:486
    oleracea, 1:418*n13*
    rapa, 3:619*n14*
    rugosa, 4:235*n1*
    saxea, 5:437

Bratta waly, 3:609, *3:610*, 3:611

Brazil Wood, 3:269*n6*, 3:270

Breadfruit, 1:343*n1*, 1:355*n1*,
    1:355*n3*

Bredo, 4:159
    mamma, 4:234

patassi, 4:513, 5:115
    tali, 4:459
    tintele, 4:511

Breynia cernua, 1:277*n22*, 3:151*n1*

Bringela, 4:169

Bromelia, 4:156
    comosa, 4:151*n1*

Bronia alba, 4:511, *4:512*, *4:513*

Broom plant, 5:86

Brooms, 4:231*n1*, 4:231*n2*

Broussonetia papyrifera, 3:38*n4*,
    3:377*n2*

Brucea amarissima, 4:265*n11*,
    5:519*n1*, 5:521*n1*

Bruful
    era conjugata, 1:454*n12*

Brufulera conjugata, 3:543*n23*

Bruguiera
    conjugata, 2:435*n22*, 2:605*n1*,
        2:607*n2*, 2:635*n2*, 3:117*n4*,
        3:271*n23*
    cylindrica, 2:605*n1*, 2:629*n1*
    sexangula, 2:607*n1*

Brungara aradua, 4:470

Brutaca, 1:473

Bryonia, 4:369*n3*, 4:451
    alba, 4:511*n2*
    dioica, 3:351*n1*, 4:445*n4*, 4:515*n2*

Bryophyllum pinnatum, 4:228*n7*

Bua sima, 3:60*n2*

Buan, 1:424
    abbal, 1:425
    tauno, 1:425, 2:493

Buckthorn, 2:183

Buckwheat, 5:25*n3*

Bud flower, 4:250

Buflossum
    lanuginosum, *3:384*, 3:385
    litoreum, 3:350, *3:380*, 3:381–
        3:383

Bugulawan, 2:5
    Ayca, 2:547

Buhula wabbal, 2:21

Buju buju, 5:65

Bulangan Tree, 2:185

Bulat, 1:365

Bulbophyllum, 5:205*n19*
  grandiflorum, 5:205*n19*
  uniflorum, 5:205*n19*

Buleyn, 1:290

Buli Buli, 2:582

Bullock's heart, 1:395*n3*, 1:397*n12*

Bullu mappa, 4:581

Bulrushes, 5:401

Bulu, *3:180*, 3:181–3:185
  baduri, 3:199
  carissa, 3:184
  casser, 3:184
  gading, 3:206
  gadu, 3:203
  gantang, 3:191, 3:197
  Java, 3:189, 3:191, 3:192, 3:193,
    3:205, 3:206
  key, 3:205
  kutsjing, 5:555
  matta kees, 5:555
  potong, 3:195, 3:199
  sammet, 3:191, 3:197, 3:206
  sera, 3:201, 3:205, 3:207, 3:209
  scru, 3:187, 3:188, 3:189, 3:193
  swangi, 1:51, 3:188, 3:189, 3:191,
    3:192, 3:193, 3:195, 3:199,
    3:202, 3:203, 3:205, 3:206,
    3:207, 3:272, 3:642, 5:21–5:22
  tombor, 3:205
  totoan, 3:192
  tsjatjar, 3:207
  tuy, 3:189, 3:190, 3:669
  wani, 3:195, 3:197
  wani besaar, 3:197

Bulun, 1:298

Bulung, 5:339

Bumit, 3:5

Bun peperon, 3:527*n8*

Buncalunga, 4:477

Bune, 3:145
  carbou, 3:145

Bunga bissu, 3:219*n14*

Bungo bungo, 5:99

Bungo plant, *5:92*, *5:96*, *5:99*

Bungum
  foemina, *5:96*, 5:99
  mas, *5:96*, 5:99

Bunius
  agrestis, 3:145, 3:147
  sativa, 3:145, *3:146*, 3:147

Bunu cutto, 3:388

Burang, 3:34

Burdock, *5:110*, 5:113–5:114

Burladora, 4:179

Burning fever leaf, *4:512*, 4:515–
  4:517

Burning Leaf, *3:174*, 3:175–3:177

Buro
  buro, 3:435
  malacco, 4:561

Burr reed, 5:38*n12*

Burseraceae, 2:227*n5*

Buru, 1:38, 1:39

Busso, 4:55

Butcher's broom, 5:311*n1*

Buto
  scytan, 5:253
  siël, 5:372

Buton tree, *3:94*, 3:95–3:96
  land, 3:97, *3:98*, *3:99*, 3:100

Butonean shrub, rope-like, *3:599*,
  3:600

Butonica, *3:94*, 3:95–3:96
  terrestris, 3:97, *3:98*, *3:99*, 3:100

Butter tree, 5:476

Butterfly Draught, 3:365

Buttu maun, 1:377

Buxaceae, 2:452*n15*, 2:479*n5*

Buxus sempervirens, 2:151*n2*,
  3:511*n2*

Buyo, 4:6
  buyo, 4:316, 4:327
  cohihu, 4:11
  lutton, 4:23
  samo, 4:317

Bythoa, 2:67

C

Caachira, 4:146

Caban cabannan, 4:270

Cabang, 1:266

Cabbage tree, 1:221*n9*

Cabbita bita, 5:567

Cabeça, 4:146

Cabi cabi, 4:168, 5:95, 5:95*n5*

Cabritten-kruid, 5:87

Cabuwo, 4:349

Cacara, 4:359, *4:390*, 4:391–4:392
  alba, 4:393, *4:394*
  bulbosa, *4:376*, 4:377–4:380
  bulu, 4:417
  djule, *4:395*, *4:396*, 4:397
  gattal, 4:419
  laut, *4:412*, 4:413–4:414
  litorea, *4:412*, 4:413–4:414
  luke, 4:395
  lungsir, 4:393, *4:394*
  marina, 4:413, 4:485
  nigra, *4:395*, *4:396*, 4:397, 4:417
  parrang, *4:386*, 4:387
  pilosa, 4:417
  pruritus, *4:418*, 4:419–4:420
  puti, 4:393, *4:394*

Cacolaa, 4:45

Cacuti cuti, 4:519

Cada Pilava, 3:55

Cadali, 3:429

Cadel avanacu, 3:347

Cadelari, 5:53

Cadelee, 4:409, 4:410

Cadelium, *4:408*, 4:409–4:410

Cadjang ticos, 4:406

Cadju, *1:468*, 1:469–1:470
  tawa, 1:491
  utan, 1:473
  wild, 1:471, *1:472*, 1:473–1:474

Cado cado, 4:331

Cadoc, 4:110

Cadoja, 5:476

Cadolini, 4:9

Caduc duc, 3:425, 3:427

Cadurpa, 4:186

Cäe, *5:162*, 5:163

Caepa silvestris, 5:306, 5:307–5:308

Caesalpinia
  crista, 2:55*n21*, 2:374*n9*,
    3:269*n6*, 3:275*n14*, 3:327*n5*,
    3:627*n1*, 4:336*n5*
  echinata, 3:269*n6*
  jayabo, 3:623*n2*
  nuga, 2:623*n4*, 3:631*n1*,
    4:494*n13*
  pulcherrima, 1:489*n2*, 3:265*n1*,
    3:467*n7*, 3:565*n2*, 3:635*n3*,
    4:141*n3*
  sappan, 2:293*n12*, 2:413*n27*,
    2:505*n11*, 2:582*n7*, 2:589*n4*,
    3:51*n5*, 3:192*n13*, 3:269*n1*,
    3:269*n6*, 3:272*n32*, 3:276*n22*,
    3:381*n4*, 3:499*n1*, 3:623*n4*,
    3:631*n3*, 5:95*n10*

Cage cage, 4:268

Cahamaisso, 5:233

Cahingora, 3:286

Caim gulur, 5:565, 5:566

Caja haba, 5:143

Cajanus cajan, 4:389*n1*, 5:281*n3*

Cajou nassa, 4:153

Caju
  Angin, 2:570
  Api Api, 2:623
  apun, 5:333
  Arang, 2:412, 2:419, 2:421
  Assam
    Besaar, 2:511
    Daun Kitsjil, 2:513
  auwaran, 4:521
  ayer, 1:385
  Backat, 2:127
  Baco, 2:582
  bandaâ, 1:349
  Bappa, 2:257
  baru, 2:332, 2:343, 3:61, 3:559
  barudan, 3:139
  Batu, 2:454
  Bawang, 2:120

belo, 2:595, 3:173
  itam, 2:595
besaar, *5:478*, 5:479
bessi, 2:440, 3:149
  benar, 2:431
  ternaats, 2:446
boaraus, 3:55
boba, 2:595, 3:69
bobata, 3:535
bodi, 3:27, 3:27*n6*, 3:30
bulian, 1:261, 1:495
buton, 3:96
Cambing, 2:205
Carbou, 2:440
Casturi, 2:63, 2:88
cauwa, 4:575
Cawan, 2:259, 2:635
ceram, 3:388, *3:396*, 3:397
corita, 1:420
Cuda, 2:363, *2:540*, 2:541–2:542
Cuning, 2:561
cutana, *5:486*, 5:487
danu, 3:71
dauwar, 5:277
Dingin, 2:337
djurang, 3:41
Dulang, 2:123
fita fita, 5:567
gabba gabba, 3:78, 3:113
galedupa, 2:8*5*, 2:86, 2:8*7* 2:88,
  3:501
Garo, 2:88
Gelam, 2:106
Gelamor, 2:259
Goa, 2:87
gorita, 1:420*n23*, 2:370, 2:577,
  3:348, 3:349–3:350
hoan, 4:405
Hollanda, 2:565
iean manassin, 4:209
Itam, 2:421, 2:422, 2:425
Jati, 2:463–2:464
Java, 2:505
Kilan, 2:109
lacca, 2:56, 2:88, 2:314, 2:316,
  2:317, 3:162, 3:251, 3:471,
  3:498, 3:499–3:502, 3:503,
  3:507, 3:511, 3:517, 3:637,
  4:507
lale, 3:567

landac, 5:511
langit, *3:148*, 3:149–3:150
lape lape, 2:552, 3:159
lapia, 3:143
laro, 5:507
lassi, 2:120, 2:481, 3:423
lingoo, 3:499
lobe, 2:549, 2:552
  daun kitsjil, 2:553
loluy, 3:336
maas, 2:499, 2:505, 3:276, 3:286
Madjannang, 2:349, 2:351
Marea, 2:323
maria, 1:387, 1:389, 2:451, 2:589
matta buta, 2:88, 2:340, 2:361,
  2:363, 3:502, 5:203
  darat, *3:410*, 3:411–3:412
mera, 1:379, 1:381, 1:392*n19*,
  2:21, 2:22, 2:543
  daun besaar, 2:548
Mcrack, 2:570
Mero, 2:309–2:311
moni, 3:513
Morea, 2:496, 2:497
moron, 3:109
morsego, 5:499
nani, 1:432, 2:429, *2:430*, 2:431–
  2:432, *2:433*, 2:434–2:435,
  2:436, 2:437, 2:505
nanni, 2:181, 3:389, *3:390*, 3:391
  besaar, *3:386*, 3:387–3:388
nenu, 3:57
nussalauw, *5:490*, 5:493
palacca, 3:127
panu, 5:485
Papua, 2:83, 2:84
Pukon, 2:631
pule, 3:524
puli, 3:571
purin, 3:289
puti, 2:102, 2:105, *2:108*, 2:109–
  2:110, 2:111, *2:112*, 2:113,
  2:259, 2:279, 3:61, 3:149,
  3:163, 3:421, 3:427, 5:159,
  5:161, 5:174, 5:189, 5:221,
  5:283, 5:555, 5:563
puttar, *5:525*, *5:526*, 5:527
Radja, 2:123, 2:389, 2:406, 2:474
rapat, 2:14, 3:517, *3:518*, 3:519–
  3:520, 3:521, 3:609

Rasamala, 2:83, 2:83*n1*, 2:84,
    2:88
russa, 3:111
Salimuri, 2:412
salowacko
    mera, 3:90
    puti, 3:87
Samarra, 2:83
sanca, 2:393, 3:395, 5:519
sappan, *3:268*, 3:269–3:272
Sarassa, 2:635
Sawa, 2:432
Sawo, 2:434, 2:435
say ambon, 5:487
Siri, *2:520, 2:521–2:522, 2:523*
Sissit, 2:519, 2:549
somin, 5:493
Sommot, 2:389
sonti, 3:387
Sore, 2:565
Sussu, 2:370
swangi, 2:635, 4:83
tanang, 3:5
tawa, 4:575
Tay, 2:181, 2:340, 2:464
Ticcos, 2:495, 2:496, 2:497
Titti, 2:469, *2:470, 2:471*
tombor, 3:45
totarra, 1:277, 3:109, 3:115
    itam, 3:113
towo, 3:626
Tsjammara, 2:570
tsjina, 2:389, 4:498
    utan, 4:491, *4:492*, 4:493–
        4:495
uda, 3:275
Ular, 2:129, *2:180*, 2:181
Ulassi, 2:481
Upas, 2:400
Wale, 2:415
Waro, 2:344
Weyer, 2:478
wuna wuna, 3:219
Cajuputi tree, 1:1
Cajut larat, 5:507
Calabassa, 4:285, *4:424*, 4:425–
    4:426
    utan, 4:426
Calacu, 1:196

Caladium
    aquatile, 1:275, 4:291, *4:292*,
        4:293
    cum liberis, 4:287
    sativum, 4:285, 4:287
    scriptum, 4:287
    spontaneum, 4:291, *4:292*, 4:293
    vicorum, 4:291
Calafur, 2:5
Calam, 5:159
Calamagrostis, 3:213, 5:26, 5:29–
    5:30, 5:31
Calambac, 2:47–2:51, 2:52, 2:53–
    2:62, 2:84, 2:88, 2:316, 2:361,
    2:367, 3:25
    Puti, 2:62, 2:63
Calambir, 1:193
Calambou, 4:333
Calamojang, 4:295
Calampoack, 1:379, 2:559
Calamum aromaticum, 4:83
Calamus
    acidus, 3:669*n21*, 5:540*n1*
    albus, 3:645*n1*
    buroensis, 3:656*n8*
    cawa, 3:659*n6*, 3:660*n11*
    equestris, 3:657*n1*, 5:10*n15*,
        5:441*n2*
    graminosus, 3:647*n7*
    pisicarps, 3:651*n3*
    viminalis, 3:655*n1*
Calangari, 4:431
Calanthe triplicata, 5:215, 5:221*n1*
Calappa, 4:429
    babi, 1:203
    bubur, 1:202
    canari, 1:202
    canarina, 1:202
    capuli-formis, 1:203
    cassomba, 1:201
    cysti-formis, 1:203
    king's, 1:203
    lansa, 1:204
    lansi
        formis, 1:204
    laut, 2:75, 5:357*n10*, *5:382*,
        5:383–5:393, 5:384, 5:397

lupa, 3:96
machaeroides, 1:202, 1:204
mera, 1:203
muda, 1:193
parrang, 1:202
pultaria, 1:202
pumila, 1:203, 4:182
radja, 1:203
regia, 1:203
royal, 1:204
rutila, 1:201
saccharina, 1:202
tubu, 1:202
vulgaris, 1:201
Calappus, 1:56, 2:11, 2:205, 2:258,
    3:7, 3:23, 3:71, 3:95, 3:300,
    3:301, 3:303, 3:305, 3:315, 3:350,
    3:453, 3:455, 3:468, 3:543, 3:571,
    3:615, 3:639, *5:178*, 5:180, 5:297,
    5:304, 5:308, 5:346, 5:473, 5:476,
    5:477, 5:503, 5:505, 5:563
  Malidiva, 1:205
  sea, 1:204, 1:205
  Sugar, 1:202
  tree, 1:189–1:191, *1:191*, 1:192–
      1:199, *1:200*, 1:201–1:219,
      1:220, 1:221, *1:222*, 1:247,
      1:253, 1:255, 1:256, 1:263,
      1:269, 1:271, *1:298*, *1:313*,
      1:414, 4:326
    potherb, 1:311, *1:312*, 1:313–
        1:314, *1:315*, *1:316*, 1:317,
        1:318, 1:319–1:320
Calawassan, 3:355
Calbahaar, 2:75, 5:359
  gunong, 5:399
  itam, 5:397
  mountain, 5:399
  puti, 5:401
Calceolaria, 5:179*n3*
Calceolus mariae, 5:179
Cale, sea, 5:170*n11*
Calembemba, 3:476
Calemojang, 4:300
Caleronkon, 2:631
Calf's foot, 4:290*n41*
Caligas Helveticas, 4:192

Callalluh, 4:160
Calleli Tree, 2:273
Callicarpa, 3:388*n11*
  cana, 3:393*n1*
  cuspidata, 3:395*n1*
  longifolia, 3:395*n1*, 3:397*n1*
Callocane, 5:339
Calmus, 2:59, 4:81, *4:82*, 4:83–*4:84*
Calodjo, 3:27
Calompon, 3:73
Caloncan, 4:381
Calophyllodendron, 2:326
Calophyllum, 2:329*n6*, 5:469*n2*
  inophyllum, 1:387*n2*, 1:392*n19*,
      2:201*n3*, 2:303*n10*, 2:321*n1*,
      2:451*n9*, 2:468*n9*, 2:522*n9*,
      2:589*n2*, 3:46*n15*, 3:95*n2*,
      3:569*n10*, 5:469*n7*, 5:471*n20*,
      5:489*n3*
  nagassarium, 5:469*n2*
  soulattri, 2:327*n1*, 2:503*n4*,
      2:525*n2*, 2:577*n2*, 3:143*n2*
Calotropis gigantea, 3:221*n33*,
    5:513*n1*
Caloway tree, 3:70, 3:71–3:72
Caltrop's herb, 4:470
Calvatia bovista, 5:251*n3*
Calyptrocalyx spicatus, 1:239*n1*
Cama
  Lilan, 2:257
  Mette, 2:242
Camacoan, 2:237
  litsjing, 2:232, 2:234
Camadjara, 4:86
  betul, 4:77
Camadu, 5:89
Camalanan, 2:237
Camari, 1:362
Camaroch, 1:362
Camarungi, 3:626
Camban malatte, 3:558
Cambang, 2:303
  cuning, *3:278*, 3:279, 4:237,
      4:238, 4:239

norotja, 4:192
  sore, 4:192
Cambodian Gum, 2:379–2:380
Cambong, 5:335
Camean, *5:490*, 5:491
Camel hay, 4:86*n7*
Camelan, 2:106, 2:195
Camelaun abbal, 4:329
Camelina, 3:335*n3*
  dentata, 2:3*n3*, 3:407*n2*
Camerarius Althaea Theophrasti,
  3:227
Camibelo, 3:27
Camiri Tree, 2:271, 2:272, 2:273
Camirium, 2:271, 2:272*i2:273*
Camiten Puti, 2:419
Cammang, 3:47
Cammar Amin, 2:250
Camolenga, 4:421, *4:422*, 4:423,
  4:441
Camomile, 2:61
Camotes, 4:370
Campana
  coerulea, 3:564
  rubra, 5:543
Campeachy Wood, 3:272*n36*,
  3:501*n24*
Campernoyle, *5:246*, 5:247–5:248.
  See also Mushrooms
  boll-like, 5:251
  devil's, 5:253
  fiery, 5:249
  land, 5:443
  stony, *5:442*, 5:443–5:444
  tree, 5:245
Campeu, 5:19
Camphor tree, 5:585–5:588
Camsia, 4:95
Camubut, 3:539
Camul, 4:320
Camulu hahua, 3:527
Camuneng, 3:511, *3:512*, 3:513–
  3:514, *3:515*, 3:516

batu, 3:514
  Chinese, 3:514
  japon, 3:516
  java, 3:514
  shrub, 3:511, *3:512*, 3:513–3:514,
    3:515, 3:516
Camuning, 5:537
Camunium, 3:511, *3:512*, 3:513–
  3:514, *3:515*, 3:516
  japonense, 3:511*n1*
  vulgare, 3:511*n1*
Cana
  Chinese, 2:228
  clove, 2:228
  tenglio, 2:228
Cananga, 2:294, 2:295–2:97, 2:295–
  2:297, 3:139, 3:149, 3:559
  odorata, 2:295*n1*, 2:301*n2*,
    2:303*n10*, 2:422*n6*
  silvestris, 2:298, 2:299–2:300
    secunda, 2:300
    tertia, 2:300
    trifolia, 2:298, 2:299–2:300
  Utan, 2:299, 2:300
  wild, *2:298*, 2:299–2:300
Canangium odoratum, 1:459*n2*,
  2:203*n5*, 2:253*n2*, 3:149*n6*,
  3:559*n13*, 5:195*n4*
Canara oleous, 2:243, *2:244*,
  2:245–2:247
Canari, 2:239, *2:240*–2:241, 2:242,
  3:67, 3:71, 3:125, 3:571
  Baggla, 2:220
  Barat, 2:223, 2:229, 2:233
  Besaar, 2:220, *2:248*, 2:249,
    2:250
  callapus, 1:202
  Chinese, 2:228
  common, 2:215, *2:216*–2:217,
    2:218–2:222
  Dammar, 2:227
  first wild, 2:223, *2:224*–2:225,
    2:226–2:228
  Kitsjil, 2:220, 2:226
  largest, *2:248*, 2:249–2:251
  laut, 1:466
  Negri, 2:215, *2:216*–2:217
  Panjang, 2:215

Pende, 2:215
scented, 2:232, 2:233–2:234,
2:235, 2:236–2:237
second wild, 2:229, 2:230, 2:231
Utan, 2:226, 2:231
Xula, 2:250

Canariformis, 2:43, 2:45

Canariopsis paucijuga, 2:233*n3*

Canaris, 2:11

Canarium, 3:569*n1*, 5:199*n14*
acutifolium, 2:239*n1*
album, 2:228*n13*
balsamiferum, 1:431*n8*, 2:233*n1*
commune, 1:202*n12*, 1:331*n2*,
1:433, 1:451*n3*, 2:11*n1*,
2:85*n7*, 2:119*n6*, 2:201*n2*,
2:215*n1*, 2:234*n5*, 2:311*n18*,
3:61*n3*, 3:125*n3*, 3:375*n6*,
3:437*n4*, 3:453*n2*, 4:14*n10*,
5:6*n23*, 5:15*n11*, 5:149*n2*,
5:173*n6*, 5:181*n42*, 5:219*n8*,
5:268*n7*
comune, 4:290*n36*
decumanum, 2:248, 2:249–2:251,
2:249*n1*, 2:270, 3:67, 3:67*n2*
hirsutum, 2:81*n16*, 2:228*n14*,
2:234*n6*, 2:236
Javanum, 2:271
legitimum, 2:231*n7*
microcarpum, 2:83*n1*
montanum, 2:231
nanarium, 2:231
nigra, 2:239, 2:240 2:241, 2:242
odoriferum, 2:245, 2:246
hirsutum, 2:232, 2:233–2:237
leve, 2:232, 2:233–2:237
oleosum, 2:82*n20*, 2:83*n1*,
2:85*n8*, 2:87*n20*, 2:88*n33*,
2:88*n36*, 2:205*n6*, 2:236*n17*,
2:243*n1*, 4:62*n30*
pimela, 2:228*n10*, 2:228*n12*
silvestre, 2:229, 2:229*n1*, 2:230,
2:231, 2:555*n4*
sinense, 2:228
trigonum, 2:81
odoratum, 2:228
vulgare, 2:215, 2:216–2:217,
2:218–2:222

zephryinum, 1:497*n3*, 2:102*n14*,
2:223, 2:224–2:225, 2:226–
2:228, 2:229*n2*, 2:522*n8*
zeylanicum, 2:227*n9*

Canaru, 4:160

Canavalia
gladiata, 4:387*n1*
lineata, 4:413*n1*, 4:414*n4*
microcarpa, 4:413*n1*, 4:414*n4*,
4:485*n5*
turgida, 4:414*n4*

Canaya, 4:14
puti, 4:30*n16*

Canckals, 2:468

Cancong, 4:461, 4:463
ajer, 4:463
utan, 4:487, 5:40, 5:43–5:45

Candarussa cuning, 3:286

Candelarium, 2:603

Canjalut, 4:366
itam, 4:366

Canjang Ya, 2:571

Canna, 5:552
de bengala, 3:637, 3:638, 3:639–
3:642, 3:642
East-Indian, 3:216
fistula, 2:123, 2:125, 2:127, 2:128,
2:129–2:130, 2:309, 2:313,
2:314, 2:495, 2:563, 2:612,
3:127, 3:265, 3:279, 4:237
indica, 3:215*n4*, 4:29*n10*,
4:63*n47*, 4:79*n1*
marshy, 3:208, 3:209
palustris, 3:208, 3:209

Cannabis
indica, 4:123, 4:124, 4:125–4:127
sativa, 4:125*n6*

Cannacora, 2:380, 3:215

Cannacorus, 4:76, 4:79–4:80

Cannacur, 4:326

Canneka de mato, 3:674

Cannelli itti canni, 3:524

Cano cano, 3:208, 3:209, 5:26,
5:29–5:30

Canre ular, 4:454

Canrular, 4:511

Cantang, 4:481

Cantharifera, 3:673–3:676

Cantjulikan, 3:61, *3:62,* 3:63

Capalla
  cassomba, 4:138
  negry, 3:6

Capamava, 1:470

Capas, *3:232,* 3:233–3:237
  antu, *3:240,* 3:241, 3:244
  besaar, *3:238,* 3:239

Capawaarita Tree, 2:296

Capaya Tree, 2:276

Capernoyle, 5:233n1

Capey
  besaar, 5:141
  gorita, 5:141, *5:142,* 5:143
  tullis, 5:141

Caphar, 4:61

Capialun, 4:515

Capillus
  nympharum, 5:171, 5:337–5:338
  veneris, 5:140
    amboinicus, *5:144,* 5:145

Capita gauha, 5:514

Capo capo, 5:333

Capock tree, 1:335, 1:335n6,
  1:501–1:503, 3:571

Capox, 3:73, 3:77

Cappa schora, 4:426

Capraria, 2:205–2:206, 5:87

Caprificus, 3:37
  Amboinica
    angustifolia, 3:34, 3:35
    aspera, 3:41
    latifolia, 3:34, 3:35
  angustifolia, 3:41
  aspera, 3:118, 5:493
  chartaria, *3:32,* 3:33–3:36
  esculenta, *3:32,* 3:33–3:36
  riparia, 3:43
  viridis, *3:42,* 3:43

Capsicum, 2:541n3, 3:547n6,
  5:346n10, 5:346n17
  agreste, 5:101

annuum, 4:183n9, 4:186n18,
  4:186n29, 5:99n4
frutescens, 1:236n43, 1:328n37,
  3:150, 3:150n11, 3:295n3,
  4:185n14, 4:186n18, 4:373n4,
  5:326n5
Indicum, 1:236
majus rubrum, 4:183
minus
  flavum, 4:185
  rubrum, 4:185
occidentale, 4:186
orientale, 4:183, *4:184,* 4:185–
  4:189
silvestre, *3:416,* 3:417, 4:564,
  5:101

Capulaga, 4:45

Capur baros, 5:586, 5:588

Cara
  mandyn, 3:425
  veela, 4:234

Carabeli, 1:362

Carambau dinar, 3:331

Carambola, 1:362, 1:363, 1:365n1

Carambu, 5:94

Caranasci, *4:530,* 4:533–4:535

Caranassi, 5:129

Carandas, 5:569

Carandje, 5:569

Carando, 4:351

Caraneang, 2:195

Carang, 5:357, 5:446
  alea, 5:430, 5:433, 5:447
  ayer, 5:449
  bonga, 5:357, 5:430, 5:432, 5:435
  carang, 5:150
  carussa, 5:444
  culat, 5:444
  gula, 5:435
  mera, 5:421
  panthey, 5:430
  rede, 5:441
  uby, 5:446

Caranna, 2:229

Caranosi, 3:258

Caranotschi, 3:293

Caraway, 4:219–4:220

Carbanga, 4:67

Carbonaria, 2:498, 2:499–2:500
  altera, 2:502
  asper, 3:429

Carcapuli, 2:326, 2:380, 4:140

Carcum, 4:60

Cardamomum, 1:234, 4:365, 5:220,
  5:261, 5:262, 5:263
  Ambonese, 4:41
  graecum, 4:45
  medium rotundum, 4:43, *4:44,*
    4:45
  rotundum, 4:247
  small, 4:43, *4:44,* 4:45
  utan, 4:41

Cardinal's flower, *4:462,* 4:465–
  4:467

Cardiopteris moluccana, 4:569n1

Cardiospermum halicacabum,
  3:595n6, 5:115n10

Carduus
  benedictus, 3:245, 3:247, 3:309,
    3:309n1, 4:135
  peregrinus, 3:332

Carelu vegon, 4:560

Carendang, 5:569

Careum indicum, 4:219–4:220

Carex, 5:33, 5:33n3, 5:33n6,
  5:223n17
  amboinica, 5:32, 5:36, 5:37–5:39
  antiquorum, 5:38
  arborea, 5:39
  culmaris, 5:39
  laevis
    major, 5:38
    minor, 5:38, 5:39

Cari Capusi, 2:337

Cari villandi, 4:495

Carica, 1:419
  papaya, 1:414n1, 3:220n21

Carim tumba, 5:77

Carimpana, 1:256, 1:261

Caringejo, 4:413

Carissa carandas, 5:191n6, 5:539n3,
  5:569n1

Carophyllum, 2:5

Carpussa batu, 5:444

Carthamus
  indicus, 1:202, *4:132,* 4:135–
    4:140
  tinctorius, 1:337n14, 2:151n6,
    2:273n9, 3:25n18, 3:52n8,
    3:75n9, 4:123n5, 4:135n3,
    4:136n10, 4:168n9, 4:263n3,
    4:282, 4:528n12, 5:58n6

Carum, 4:219–4:220

Carunrun, 1:439

Carvos, 2:5

Caryophyllaster, 2:549, 2:553
  alter, 2:22, 3:160, 3:161–3:162
  litoreus, 3:368, 3:369, 3:393
  primus, 3:401
  terrestris, 2:22

Caryophyllis
  aromaticus, 2:3n1, 2:7, 2:19
  Plinii, 2:7
  silvestre, 2:20, 2:21–2:22

Caryophyllum, 3:275n19, 4:45
  saxeum, *5:436,* 5:439–5:440

Caryota, 1:282
  rumphiana, 1:56, 1:242n14,
    1:281n1, 1:281n3, 3:443n6

Caseolare, 2:582

Casimbucan, 4:489

Cassava, 4:371n8

Casse selan, 5:305, 5:309

Cassia
  alata, 5:531n1, 5:561n7
  aromatica, 2:123
  caryophyllata, 2:98
  cinnamomum, 2:60
  fistula, 1:57, 2:121, 2:122, 2:123–
    2:126, 2:126, 2:134, 2:309,
    2:309n5, 2:313n3, 2:410,
    2:474, 2:495n2, 2:612n3,
    3:127n5, 3:265n2, 3:279n2,
    3:589n7, 5:507n2, 5:561
  indica, 2:130

silvestris, 2:127, 2:128, 2:129–2:130
glauca, 3:279*n1*, 4:237*n1*
javanica, 2:127*n1*, 2:309*n6*, 2:563*n2*, 4:237*n5*, 4:513*n16*
lignea, 2:95, 2:96, 2:102
mimosoides, 5:283*n1*
occidentalis, 2:309*n9*, 3:87*n6*, 3:279*n3*, 4:237*n2*
solutiva, 2:123
sophera, 3:267*n21*
tora, 3:87*n6*, 3:267*n24*, 3:279*n3*, 4:237*n6*
Cassia Flower, 3:254*n41*
Cassomba, 4:136
Cassomba Kling Tree, 2:114, 2:115–2:117
Cassutha cornea, 5:557
Cassutha plant, 3:315*n6*
Cassuvium, 1:469
silvestre, 1:473, 3:297
Cassytha filiformis, 3:315*n6*, 4:583*n1*
Casta, 3:31
Castilian flower, 4:247*n7*
Castor oil plant, 3:335In2
Castoreum, 3:243
Castori, 3:243
Casuarina
equisetifolia, 2:569*n1*, 5:208n10, 5:379*n2*
litorea, 2:568, 2:569–2:571, 2:572, 2:573
montana, 2:106, 2:568, 2:569–2:571, 2:572, 2:573
Casuarina tree, 5:379
Cat mint, 4:257, 4:257*n3*, 4:257*n6*, 4:258, 4:377*n3*
Cat's hair, 5:555
Cata
picinna, 4:439
putia, 5:491
Catappa
domestica, 1:466
litorea, 1:466
silvestris, 1:466

Catappa tree, *1:464*, 1:465–1:467
beach, 1:465, 1:466
forest, 1:466
Catappan, 3:61, 3:65, 3:95, 3:169
laut, 1:466
utan, 1:466
Cataputia, 3:283
major, 3:336
minor, 2:359
Catecommer, 4:221
Cateppeng, 3:279
Cathartica, 2:123
Catilang, 5:520
Catimban, 5:267, 5:273, 5:274
besaar, 5:274
durium, 5:274
kitsjil, 5:274
Catjang, 3:103, 3:137, 3:560, 5:521
djintam, 4:411
tsjina, 5:199
Catnip, 4:377*n3*
Catsja, 1:234
Catsjem tsjem, 1:439
Catsjopiri, 5:517
Catsju, 3:123
piri, 3:289, 3:327
utan, 3:327
Catta Mala, 3:91
Cattagasturi, 3:244
Cattail, 1:256*n8*, 1:350*n5*, 1:351, 3:215, *3:318*, 3:319–3:320
wild, 3:321, *3:322*, 3:323
Cattam cattam utan, 4:487
Cattammon, 1:379
Cattan, 4:113
Cattappa de mato, 3:590, 3:591
Cattari, 5:219
Catti catti, 3:145, 3:623, *3:624*, 3:625–3:626, 3:627, 4:494
kitsjil, 3:631, 3:635–3:636
pante, 3:631, *3:632*, 3:633
Catti Catti Garo Tree, 2:195
Catti marus, 3:91, *3:92*, 3:93, 3:115, 4:326

Cattoen, 3:234

Catu
  paerou, 4:403
  paeru, 4:413
  picinna, 4:446
  tsjambu, 1:381

Catulampa, 3:61, *3:62*, 3:63

Catundeng, 2:452

Cauda felis, 3:38, *3:318,* 3:319–
  3:320, 3:323
  agrestis, 3:321, *3:322,* 3:323

Caunga, 1:230

Cautschoa, 5:531

Cavalam, 3:78

Cawo cawo, 1:502

Caylupa, 1:502

Cayman's tongue, 4:221

Cayman's ubi, 4:343

Cayman's Wood, 3:329

Caymoni, 3:513

Cebar, 4:221

Cedar Tree, 2:569*n2*

Cedrela febrifuga, 2:643*n5,* 2:643*n6*

Ceiba pentandra, 1:501*n1,* 3:73*n2,*
  3:237*n46,* 3:569*n2,* 5:495*n5*

Cello Laut, 2:496

Celosia
  argentea, 5:49*n3*
  cristata, 4:115*n2*

Celtis
  cinnamomea, 2:181*n16,* 2:463*n4*
  philippensis, 2:521*n1,* 2:522*n13,*
    3:553*n2*

Cenau, 2:344

Cenjorins, 4:6, 4:11

Centaurea benedicta, 3:245*n2,*
  4:135*n5*

Centella asiatica, 1:55, 3:311*n8,*
  4:472*n21,* 4:513*n13,* 4:523*n1,*
  5:5*n17,* 5:66*n11,* 5:72*n16,*
  5:80*n14*

Cephalanthera, 5:179*n2*

Ceram, 1:38, 1:48

Ceramese shrub, *3:396,* 3:397

Cerastium, 5:43*n1*

Ceratium, 3:496, 3:497

Ceratophyllum submersum, 5:339*n5*

Ceratopteris thalictroides, 4:465*n2,*
  5:331*n1*

Cerbera, 2:371
  lactaria, 1:420*n23*
  manghas, 1:327*n25,* 2:279*n6,*
    2:369*n1,* 2:385*n21,* 2:391*n5,*
    2:577*n5*

Cerberi valli, 4:516

Ceriops tagal, 2:629*n8,* 2:629*n10*

Chabazizi, 5:13

Chacani, 1:226

Chadira, 3:507

Chadja, 2:72

Chaetomorpha javanica, 5:170*n10,*
  5:337*n3*

Chalungan, 4:35

Chama
  crista pavoniana Brisiliana, 5:283
  decumana, 5:399

Chamaebalanus japonica, 4:468,
  4:473–4:474

Chamaecissus, 4:527

Chamaemelum nobile, 4:199*n4*

Chamaemorus indica, 3:323

Chamaerops, 1:280

Chambasal, 1:348

Chamelea, 2:361

Chamomile, 3:257*n8,* *4:198,*
  4:199–4:201, 4:265*n9*
  Chinese, 4:200*n11*
  German, 4:199*n2*
  wild, 4:199*n2*

Champaca flower, 5:196*n18*

Champacca Tree, 2:303, 2:305

Champadaha, 1:348

Chamuta, 1:273

Chandana, 2:67

Charcoal shrub
  first, *3:400,* 3:401–3:402
  rough, 3:403

Charcoal Tree, 2:*498*, 2:499–2:500

Charitoblepharon, 5:373

Chartamus indicus, 3:75

Chaste Tree, 3:255*n2*

Cheilanthes tenuifolia, 5:140*n10*

Chelidonium, 4:59*n9*

Chenopodium quinoa, 4:159*n17*

Cheramela, 5:526, 5:529

Cherry
  Indian overseas, 5:115*n1*
  winter, 5:115*n1*, 5:117

Cherry tree, 2:3
  Ceylonese, 3:359
  Peacock, 3:359

Cherry-Shaped shrub, 3:*418*, 3:419

Chervil, 5:71, 5:71*n4*

Chestnut, 5:353, 5:354, 5:473
  earth, 5:235, 5:235*n29*

Cheyro, 4:213

Chickweed, 5:43*n1*

Chicoreus ramosus, 2:87*n26*

Chili shrub, wild, 3:*416*, 3:417

Chilocarpus, 3:524*n10*, 3:549*n11*

China root, 4:501–4:502, 5:235*n27*

Chincapalones, 4:6, 4:6*n27*, 4:9

Chineesche plum, 2:437

Chinese arbor calappoides, 1:320

Chinese Ash tree, 3:96*n12*

Chinese Flower, 3:266

Chinese leaves, 4:528

Chinese Rose-boom, 3:224

Chinomieha, 4:431

Chirotheca marina, 5:*452*, 5:455

Cho, 3:193

Chribaz, 4:431

Chrysanthemum indicum, 4:199*n10*,
  4:470*n9*

Chrysopogan acidulatus, 5:23*n1*

Chu-chin, 3:219*n15*

Cibotium barometz, 5:130*n19*

Cicadaria, 3:*158*, 3:159

Cidra Tree, 2:143

Cijavo kitsjil, 1:411

Cingulum
  petrae, 4:528
  terrae, 5:164, 5:165

Cinnamomum, 5:583*n1*
  camphora, 3:253*n30*, 5:432*n18*,
    5:585*n1*, 5:585*n2*
  cassia, 2:93*n3*
  culitlawan, 1:353*n18*, 1:391*n14*,
    1:453*n8*, 1:486*n9*, 2:93*n1*,
    3:118*n21*, 3:415*n8*, 3:503*n3*,
    4:63*n39*, 4:225*n25*, 4:262*n9*,
    4:325*n5*, 5:313*n11*, 5:483*n2*,
    5:575*n3*, 5:583*n2*, 5:584*n3*
  javanicum, 2:99*n1*
  xanthoneurum, 2:95*n5*

Cinnamon, 1:42, 2:59, 2:93, 2:95,
  2:97

Cirrhipathes
  anguina, 5:369*n2*
  rumphii, 5:369*n1*
  spiralis, 5:369*n3*

Cirsion, 4:195

Cirsium, 4:195*n3*

Cissus, 4:561*n2*
  aristata, 4:507*n10*, 4:510*n19*,
    4:561*n1*, 5:225*n24*
  quadrangularis, 3:611*n11*
  repens, 3:611*n12*, 4:507*n1*,
    4:561*n7*

Cistus, 5:87

Citroenen Tree, 2:143

Citrullus, 4:432
  colocynthis, 3:339*n19*, 4:426*n13*
  lanatus, 4:431*n1*

Citrus, 1:337*n14*, 2:311*n17*, 2:609,
  3:65*n2*, 3:321*n5*, 3:557*n2*,
  3:569*n3*, 4:478*n14*, 5:539*n2*
  abversa, 3:118*n15*
  antiquorum, 2:145
  aurantifolia, 1:379*n11*, 1:395*n2*,
    2:155*n1*, 2:195*n2*, 2:209*n8*,
    2:428*n12*, 2:435*n22*, 2:467*n2*,
    2:513*n4*, 3:105*n4*, 3:573*n3*,
    3:607*n3*, 3:619*n12*, 4:138*n40*,

4:275*n3*, 4:291*n3*, 4:347*n2*,
5:326*n9*, 5:415*n9*, 5:493*n5*,
5:531*n9*
  aurantium, 1:433, 1:451*n2*,
    2:141*n12*, 2:163, 2:164, 2:165,
    2:167*n1*, 2:427*n4*, 2:467*n3*,
    3:55*n6*, 4:287*n13*, 5:569*n2*
  bergamia, 2:152*n8*
  chinensis, 5:281*n13*
  hystrix, 1:509*n2*, 2:141*n13*,
    2:147*n1*, 2:148*n8*, 2:149*n11*,
    2:152*n8*, 3:275*n16*, 4:138*n40*,
    5:304*n30*
  japonica, 2:162*n4*, 5:484*n1*
  maxima, 1:325*n12*, 1:341*n38*,
    1:392*n20*, 2:139*n5*, 2:517*n2*
  medica, 2:143*n1*, 4:6*n24*
  nobilis, 1:509*n3*, 2:263*n3*
  obversa, 2:153, 2:154, 3:134*n8*,
    3:192*n14*, 3:205*n8*, 3:535*n7*,
    3:543*n20*, 4:138*n40*,
    4:269*n22*, 4:445*n8*, 4:460*n11*,
    5:53*n24*, 5:101*n6*, 5:165*n6*
Clausena, 3:142*n12*
Cleaving leaf, 5:114
Cleisostoma subulatum, 5:202
Clematis, 4:520, 4:521–4:522
Clematiti
  samboinica, 4:556
Clematitis, 4:555
  indica, 4:556
  ronica, 4:557, 4:558, 4:559–4:560
Clerodendrum, 3:560*n20*, 5:259*n3*
  buchananii, 3:365*n1*
  commersonii, 3:560*n25*, 3:617*n3*,
    3:619*n12*
  rumphianum, 3:363*n1*
  speciosissimum, 3:365*n1*, 5:259*n2*
  squamatum, 3:363*n1*
Clitoria ternatea, 3:563*n8*, 5:281*n4*
Clitoris principissae, 3:563
Clompan tree
  large
    3:73, 3:74, 3:75
  small, 3:76, 3:77–3:79

Clompang
  burong, 3:78, 3:79
  tsjendab, 3:78
Clompant Kitsjil, 3:78
Clompanus, 3:79
  funicularis, 3:587
  major, 3:73, 3:74, 3:75, 4:137
  minor, 3:76, 3:77–3:79, 3:113,
    4:321
Clove, 1:39, 1:41, 1:42, 1:43, 1:44,
  1:46, 1:47, 1:48, 1:61, 1:99,
  1:234, 1:369, 1:369*n3*, 1:387,
  3:23, 3:103, 3:108, 3:141, 3:255,
  3:275*n19*, 3:517, 3:521, 3:569,
  4:63, 4:183, 5:208, 5:222*n12*,
  5:436, 5:439–5:440
  Garioffel, 3:275
  red, wild, 3:160, 3:161–3:162
  small-leaved, 3:161
Clove Tree, 2:2, 2:3–2:9, 2:287
  culture and harvesting, 2:11–
    2:16
  King's Cloves, 2:5
  Royal, 2:17, 2:18, 2:19
  wild, 2:5, 2:11, 2:20, 2:21–2:22
Clusius, 4:527
Clypearia
  maritima, 3:135, 3:277
  rubra, 3:135
Cnicus, 4:132, 4:135–4:140
Coä, 4:373, 4:379
Coal Wood, 2:412
Coäla, 1:265
Coauo das ilhas, 5:395
Coäve, 1:327
Cobe, 4:511
Coccinea cordifolia, 4:511*n1*,
  4:513*n10*, 5:115*n8*
Cociaeum variegatum, 5:58*n4*
Cockle-burr, 5:71*n2*
Coco-de-mer, 1:204*n32*, 5:357*n10*
Cocoja, 3:459
Coconut, 1:371, 1:414*n2*, 2:11*n3*,
  2:327, 2:575, 3:95*n7*, 3:111*n5*,
  3:118*n13*, 3:181*n1*, 3:247*n12*,

3:433*n2*, 3:442*n4*, 3:571*n19*,
3:637*n1*, 4:182*n36*, 4:290*n37*,
4:361*n10*, 4:460*n9*, 4:513*n15*,
5:17*n5*, 5:563*n2*
dwarf, 1:202*n21*
island, 5:395*n5*
King, 1:202*n24*
Maldivian, 5:382, 5:383–5:393
milk, 4:460*n10*
palm, 4:25*n2*, 4:123*n3*,
4:137*n22*, 4:267*n3*
pinang, 1:226*n29*
sea, 5:357*n10*
wild potherb, 1:313*n6*
Cocos de Maldiva, 1:204, 5:382,
5:383–5:393
Cocos nucifera, 2:258*n18*, 3:219*n20*,
4:182*n36*, 4:268*n13*, 4:326*n8*
Cocos tree, 1:189–1:191, *1:191*,
1:192–1:199
Coculus Indi, 3:150, *3:526*, 3:527–
3:529, *3:528*, *3:529*
Cocus, 1:196, *1:197*
indica, 1:215
maldivicus, 5:383–5:393
minor, *5:394*, 5:395–5:397
melindanus, 5:384
Cocx-Indi, 1:196
Codda panna, 1:249, 1:250, 1:300
Codi calli, 5:581
Codiaeum, 3:283, *3:284*, 3:285–
3:287
bractiferum, 3:291*n1*
chrysosticon
angustifolium, 3:285
larifolium, 3:285
medium, 3:283, 3:285
erythrosticton
parvifolium, 3:285
nigrum
medium, 3:285
minus, 3:286
parvifolium
viride, 3:286
simplex, 3:283, 3:283*n1*
sylvestre
taeniosum, 3:283, *3:288*, 3:289
variegatum, 2:489*n3*, 2:503*n10*,

3:283*n1*, 3:289*n1*, 4:168*n11*,
5:495*n2*, 5:503*n3*
Codiho, 3:283, *3:284*, 3:285–3:287,
3:314, 4:168
daun kitsjil, 3:289
Papua, 3:289
ribbon-like, *3:288*, 3:289
small-leaved, 3:289*n3*
tsjina, 5:495
utan
wild, *3:290*, 3:291
Codja janti, 3:282
Codong seruni, 4:470
Coeloglossum bracteatum, 5:180*n16*
Coelogyne
celebensis, 5:201*n6*
rumphii, 5:201*n6*
Cofassus, *2:450*, 2:451–2:453,
3:513, 5:231
citrine, 2:454, 2:455
Gabba Gabba, 2:452
Lacki Lacki, 2:452
Parampuah, 2:452
yellow, 2:454, 2:455
Coffo, 4:25
Cohebba magolo tsjiffi, 3:349,
3:577
Cohobba, 4:181
Coix lachryma jobi, 1:244*n21*,
4:101*n1*, 4:101*n2*, 5:41*n1*
Cojo, 4:6
coratsje, 4:10
cossi malauw, 4:9
Cokos, 1:196
Colaka, 2:424
Colchicum, 5:309, 5:309*n6*, 5:310
Colendang, 1:485
Coletta veëtla, 5:313, 5:314
Coleus, 4:257*n6*
amboinicus, 4:228*n9*
blumei, 4:259*n1*
tuberosus, 4:497*n3*, 5:346*n16*
Colinil, 4:146
Colla tala, 3:355
Colobbe, 4:425

Colocasia, 4:17, 4:288, 4:288*n24*, 5:321
  esculenta, 1:275*n2*, 3:81*n2*,
    3:117*n7*, 4:17*n24*, 4:285*n2*,
    4:285*n4*, 4:291*n1*, 4:299*n2*,
    5:319*n2*, 5:326*n8*, 5:331*n2*,
    5:333*n3*, 5:335*n4*
  indica, 4:277*n5*, 4:277*n6*
  moderna, 4:288
  neotericorum, 4:288, 5:322
  veterum, 4:288, 5:322

Coloditi
  manoora, 4:523
  masou, 4:523

Colombine, 5:213*n7*

Colopale, 4:309

Colosino, 4:535

Colot, 4:363

Colubrina
  arborescens, 5:507*n6*, 5:507*n8*
  asiatica, 3:595*n1*

Columbia subobovata, 3:113*n1*,
  3:115*n1*

Columella
  geniculata, 4:510*n13*
  trifolia, 4:510*n15*, 4:515*n1*

Comangi, 4:211
  utan, 4:209

Combang merak, 3:266

Combili, 4:352, 4:353–4:355, 5:213
  bulu, 4:353
  fansuri, 4:353
  mera, 4:354
  tsjampadaba, 4:354

Combwood, 2:519

Comfrey, 5:57, 5:57*n1*

Comi Comi, 2:561, 3:51

Cominsia, 4:29*n8*
  gigantea, 4:28, 4:30*n15*,
    4:579*n10*
  rubra, 4:27*n3*, 4:29*n9*

Commelina, 5:175*n3*
  benghalensis, 5:44*n15*, 5:175*n4*
  nudiflora, 5:175*n2*
  obtusifolia, 5:44*n18*

Commersonia bartramia, 1:56,
  1:209*n3*, 1:277*n20*, 2:597*n2*,
  3:109*n1*, 5:174*n8*

Compar mangas, 5:397

Condang, 3:34
  casse, 5:574, 5:577

Condondong, 1:379, 1:439, *1:440*,
  1:441, 2:385, 2:528
  Malacca, 1:443
  Malaccan, *1:442*, 1:443

Condondum, 1:439
  malaccense, 1:443

Condong Waru, 2:332

Condor, 4:421

Condorius, 3:84, 3:567

Conium maculatum, 2:71*n1*

Connarus, 3:587*n1*

Conocephalus amboinensis, 3:582*n14*

Consily de greyn, 4:257

Contsja, 2:561

Convallaria majalis, 5:211*n5*

Convolvulus, 3:564*n16*, 4:507
  caeruleus, 4:483
  foetidus, *4:488*, 4:489–4:490
  lacvis
    indicus, 4:491
      major, 4:475, *4:476*, 4:477–
        4:478
      minor, 4:479, *4:480*, 4:481
  marinus, 4:484, 4:485–4:486
  minor, 4:191
  riparius, 4:484, 4:487, 4:487*n1*

Conyza
  cadaverum, 5:107, 5:108
  indica
    mas, 5:108
    minor, 5:107, 5:108
  malabarica, 5:109
  mas, 5:107, 5:108
  odorata, 5:106, 5:107–5:109,
    5:107*n1*, 5:108

Copa puti, 3:558

Copaal Tree, 2:210

Coparay, 3:219

Copernicia australis, 2:229*n5*

Coph china, 4:497

Copher, 3:250

Coppa Puckuri, 2:303

Coppanabbat Tree, 2:299

Coppatum Tree, 2:303

Copra, 1:196

Coquinho, 5:384, 5:395

Coquo de Maldiva, 5:384

Coquos de melinde, 5:395

Cor Sancti Thomas, 3:476

Coral
  bastard, 5:424
  bead, 5:357*n3*
  black, 5:361, *5:362*, 5:363–5:368
  blood, 5:439
  ginger-shaped stone, 5:430
  orange, 5:358*n13*
  red, 5:359, 5:401
  red bastard, 5:416
  shrubs, 5:357
  sugar, 5:435*n4*
  sweetmeat, 5:435
  white, 5:419

Coral Pea, 3:565*n1*

Coral tree, 3:329, 3:563*n3*

Coral-bearing tree, *3:82*

Coralariam
  latifoliam, *3:82*, 3:83–3:84
  parvifoliam, *3:82*, 3:83–3:84

Corallaria, 3:567
  latifolia, 3:85, *3:86*
  parvifolia, 2:310, 3:90, 3:563,
    3:567, 3:568

Corallis alga, 5:340

Corallium
  album indicum, 5:414
  nigrum indicum, 5:361, *5:362*,
    5:363–5:368
  rubrum, *5:359*

Corallodendron, 2:351

Corchorus
  capsularis, 5:69*n5*
  olitorius, 4:131*n1*

Corda quadrocando, 3:611

Cordex foetidus, *5:486*, 5:487

Cordia
  myxa, 3:47*n1*, 3:49*n12*, 3:65*n3*,
    3:103*n2*, 3:154*n6*
  subcordata, 2:332*n2*, 2:343*n1*,
    2:412*n17*, 2:428*n8*, 2:437*n28*,
    2:449*n7*, 2:495*n3*, 2:584*n14*,
    3:71*n2*, 3:630*n13*

Cordumeni, 4:45

Cordyline fruticosa, 3:313*n1*,
  5:541*n3*

Core som, 5:548

Coriander, 4:410, 5:77, 5:511, 5:555

Coriandrum sativum, 1:298*n29*,
  3:250*n8*

Corius, 2:493

Cornos dos Diabos, 2:542

Corona ariadnes, 4:539, *4:540*, 4:541

Corona Borealis, 4:539*n4*

Corpinus, 5:242*n13*

Cortex
  acris, 5:481
  caryophylloïdes, 2:93, 2:94, 2:95–
    2:98
  consolidans, 3:517, *3:518*, 3:519–
    3:520
  filarius, 5:489
  igneus, 2:102, 5:481, *5:482*,
    5:483–5:484
  oninius, 2:89–2:92
  papetarius, 3:163, *3:164*
  piscatorum, *3:398*, 3:399
  saponarius, 2:200, 3:413, *3:414*,
    3:415

Cortomi, 5:557

Corypha, 1:247–1:251
  elata, 1:266*n10*
  gebanga, 1:265*n4*, 1:266*n10*,
    1:266*n13*
  umbraculifera, 1:265*n2*
  utan, 1:265*n2*, 1:266*n10*,
    1:266*n13*, 1:310*n12*

Cosmosandalos, 5:183

Cossir, 3:387

Costus
  indicus, 5:469
  rumphianus, 5:275*n3*, 5:275*n7*,
    5:559*n1*

speciosus, 2:87*n23,* 2:138*n9,*
    2:528*n9,* 3:282, 3:549*n13,*
    4:63*n40,* 5:275*n3,* 5:275*n7,*
    5:511*n10*
Cotisso marau, 5:150
Cotoarus, 2:237
Cottam, 4:253
Cotton, 3:232, 3:233–3:237
    Brasilian, 3:233
    broad-leaved, *3:238,* 3:239
    devil's, *3:240,* 3:241
    shrub, 3:233
    wild, 3:38
Cotton soccus tree, *1:350,* 1:351–
    1:353
Cottreman, 2:39
Cotyledon, 4:527
Cough herb, 5:65, 5:67, 5:71
Council tree, 3:27*n4*
Covalam, 1:506
Cow Thorn, 3:503, *3:504*
Cow-wheat, 5:46, 5:47 5:48
Cowackil, 2:577
Cowpeas, 4:383*n1,* 4:383*n4,*
    4:401*n1*
Cowslip, 3:265*n4*
Crab's Eyes, 3:565*n1*
Crab's leaf, 4:481, 4:485
Crackling plant, 4:545, *4:546*
Crataegus, 5:533*n2,* 5:533*n3*
Crateogonum amboinicum, 5:46,
    5:47–5:48
Crayebessen, 3:429
Creeping Jenny, 3:293*n2*
Crematogaster, 2:387*n4*
Crepitus lupi, 5:251
Crevel-kruid, 4:419
Cricket tree, 2:475*n1,* *3:158,* 3:159
Crickets' House, 3:427
Crinum
    asiatica, 1:56, 2:406*n6,* 4:273*n2,*
        5:123*n14,* 5:126*n14,* 5:299*n2,*
        5:301*n7,* 5:301*n10*

rumphii, 5:301*n7*
zeylanicum, 4:273*n1,* 5:299*n3*
Crissan, *5:32,* 5:36, 5:37–5:39
Crista pavonis, 3:84, *3:264,* 3:265–
    3:267, 3:269, 3:279, 3:282, 3:363,
    3:467, 3:565, 3:633
Crithmum
    indicus, 5:315, *5:316,* 5:317–
        5:318
    maritimum, 5:315*n1,* 5:317*n17*
Crocodile leaf, 3:373, *3:374,* 3:375,
    4:584*n10*
Crocodile's molar, 5:507
Crocomagma, 4:60
Crocus, 4:60*n13*
    indicus, 4:60
    saracenicus, 4:136
Crocus, autumn, 5:309*n6*
Croho
    batu, 4:10
    lacki lacki, 4:10
    parampuan, 4:10
Crotalaria, 4:230, 4:231–4:232
    chinensis, 4:232*n8,* 5:281*n11*
    linifolia, 5:281*n8,* 5:281*n9*
    montana, 5:281
    quinquefolia, 4:231*n4*
    retusa, 4:231*n3*
Croton
    tiglium, 2:593*n3,* 3:368*n4,*
        3:473*n3*
    variegatum, 3:283*n1,* 3:289*n1*
Crow Fruits, 3:429*n17*
Crust of trees, 3:613, *3:614,* 3:615–
    3:616
Crusta
    accarbarii, 5:365, 5:367
    arborum, 1:207, 3:613, *3:614,*
        3:615–3:616
    ollae, *4:530,* 4:533–4:535
    terrae, 5:129
Cryptocarya, 2:535*n2*
    massoi, 1:428*n8,* 1:486*n10,*
        2:89*n2,* 3:622*n8,* 4:63*n38,*
        5:469*n5,* 5:483*n3*
Cuban Abbal Tree, 2:300

Cuban hadji, 3:266

Cubilia cubili, 5:473*n1*

Cucumber, 4:421, 4:432
bisnagarican, 4:538
bitter, 4:445, 4:447*n1*
Butonese, 4:435
Chinese, 4:19*n6*, *4:430*, 4:435–
4:436, 4:538
maderaspatensean, 4:538
malabar, 4:443, 4:445
mouse, *4:536*, 4:537–4:538
mush, 4:437
squirting, 4:269*n17*
white, 4:435
wild, 4:445

Cucumis
anguinis indicus, 4:440, 4:441
conomon, 4:435*n3*
indicus, 4:430, 4:435–4:436
melo, 4:421*n4*, 4:435*n3*
murinus, 4:536, 4:537–4:538
sativus, 4:19*n6*, 4:435*n2*

Cucurbita
lagenaria, 4:285, *4:424*, 4:425–
4:426
pepo, 4:427*n6*
vulgaris indica, 4:427, 4:428,
4:429

Cudi pariti, 3:239

Cudran, 5:540

Cudrang, 3:505, *3:506*, 3:507–
3:508
Ambonese, 3:510
wild, 3:510

Cudrania javanensis, 5:533*n4*,
5:540*n1*

Cudranus, 3:505, *3:506*, 3:507–
3:508
ambonicus, 3:505, 3:507
bimanus, 3:505
javanensis, 3:505*n2*

Cui tick, 3:188

Cuinbang Futubawa, 3:223

Cujavillis, 1:411

Cujavo, 1:461, 3:279
Batavia, 1:404
lacki lacki, 1:411

negory, 1:404
utan, 1:407

Cujavus, 5:578, 5:579
agrestis, 1:407
domestica, 1:404

Culan, 1:481, 3:487

Culang, 1:477
besaar, 3:485

Culat
api, 5:249
batu, 5:233, 5:245
ela, 5:239
pajong, 5:242, 5:242*n11*, 5:246,
5:247–5:248
pala, 5:239
sagu, 5:239
sondoc, 5:247
tana, 5:243
tangan, 5:248
tsjampadaha, 5:237

Culebet, 4:381

Culilawan, 2:91, 4:63

Culit
Api, 2:102
api, 5:483
Calappa, 1:197
gugor, 1:420
lawan, 2:93, *2:94*, 2:95–2:98,
3:415, 5:483, 5:575, 5:583,
5:584
papeda, 3:163, *3:164*

Culivo dudi, 4:426

Culombo, 4:392

Culong Tsjutsju, 3:325

Culu Tsjutsju, 3:325

Cumbang
Cayrano, 2:296
Jatu, 2:291
Satu, 2:589
sore kitsjil, *3:228*, 3:229–3:231
waribang, 3:219

Cumbulam, 4:423

Cunil, 4:60

Cunin besar, 4:61

Cuning, 4:60
Java, 4:61

tommon, 4:61
warangan, 4:61, 4:63

Cunjet sebatta, 4:61

Cunyn, 4:60

Cupessa, 3:321
saxatilis, 3:323

Cupressopathes
abies, 5:377*n1*, 5:377*n3*
pumila, 5:377*n3*

Cupressus, 3:437*n3*
marina, 1:383, 5:377

Curang curang, 4:268

Curanga fel-terrae, 3:609*n4*

Curculigo
capitulata, 4:35*n12*, 5:215*n17*,
5:219*n1*, 5:223*n10*
orchioides, 5:223*n10*

Curcuma, 1:234, 1:234*n23*, 1:353,
2:304*n18*, 3:215*n3*, 4:49, 5:5,
5:217, 5:219, 5:310, 5:444,
5:511*n9*
aeruginosa, 4:66*n4*
agrestis, 4:60, 4:61
domestica, 1:353*n12*, 2:101*n5*,
4:534*n13*, 5:182*n52*
major, 4:57, 4:58, 4:59–4:63
minor, 4:59–4:63
heyneana, 4:66*n6*
longa, 1:353*n12*, 1:506*n13*,
2:304*n18*, 3:118*n12*, 4:57*n3*,
4:59*n9*, 4:65*n3*, 4:69*n12*,
5:53*n21*, 5:170*n6*, 5:310*n10*
mangga, 4:67*n9*
petiolata, 4:60*n11*, 4:66*n7*,
5:213*n5*
purpurascens, 4:66*n8*
silvestris, 4:63, 5:213, 5:213*n5*
viridiflora, 4:66*n8*
wild, 4:63
xanthorrhiza, 4:65*n3*
zedoaria, 2:405*n2*, 4:65*n3*,
4:66*n5*

Curry, 1:353*n12*, 4:219*n10*,
5:53*n21*, 5:170*n6*

Curutu Pala, 2:371

Cus cus, 5:497

Cusambium, 1:429

Cuscus Wrapper, 3:297

Cuscuta epilinum, 3:117*n8*

Cussambi tree, 1:429, *1:430*,
1:431–1:432

Cussu, *5:32*, 5:33–5:34

Cussutha indica, 4:580, 4:583–4:584

Custard apple, 1:395*n1*, 1:397*n12*,
3:107*n9*

Cusu cusu, 5:23, *5:24*

Cutsjubong, 4:179

Cuttlefish Leaf, *3:348*, 3:349–3:350

Cyanotis moluccana, 5:44*n18*

Cyanthilium cinereum, 5:67*n1*

Cyathea, 5:121*n1*
amboinensis, 3:615*n8*, 5:121*n1*
rumphiana, 5:122*n5*

Cyathelia
axillaris, 5:419*n1*

Cyathula prostrata, 5:49*n1*

Cycas
revoluta, 1:320*n19*, 1:320*n20*,
5:477*n1*
rumphii, 1:311*n1*, 2:489*n2*

Cyclophorus nummularifolius,
5:205*n16*

Cycloseris cyclolites, 5:444*n15*

Cylang bantam, 1:477

Cylon, 3:234

Cymbidium wallichii, 5:181*n28*

Cymbopogon
citratus, 1:276*n10*, 4:85*n2*
globosus, 4:86*n5*

Cynanchum ovalifolium, 3:598*n3*,
4:543*n2*

Cynodon dactylon, 5:6*n22*, 5:19*n1*

Cynometra
cauliflora, 1:445*n1*, 2:623*n3*
ramiflora, 1:449*n1*

Cynomorion, 1:446
silvestre, 1:449

Cyperus, 4:40, 5:15*n2*, 5:38*n8*
babylonicus, 4:33, 4:35, 5:5
dulcis, 5:11, 5:12, 5:13

floridus, 5:2, 5:3–5:7, 5:5, 5:6,
  5:15, 5:38, 5:38*n11*
indicus, 4:48, 5:5
inodorus, 5:4–5:5, 5:5
legitimus, 5:5
littoreus, 5:10
longus, 4:33, 5:8, 5:9–5:10,
  5:11*n5*
maritimus, 5:5
mas, 5:38, 5:38*n11*
odoratus, 5:4*n4*, 5:38*n11*
rotundus, 3:311*n5*, 3:320,
  3:320*n8*, 4:311*n3*, 5:2, 5:3–5:7
    odoratus, 5:5
Cyphi maccassarense, 3:501
Cypria Theophrasti, 1:197
Cypripedium, 5:179*n3*
  calceolus, 5:183*n57*
Cyprus, *3:248*, 3:249–3:254, 4:136
Cytandra decurrens, 5:255*n1*
Cytisus, 4:231*n2*, 4:389, 4:389*n2*

D

Dabanus, 2:457–2:461
Dactyli indici, 2:133
Dactyloctenium aegypticum, 4:538*n6*,
  5:17*n7*, 5:35*n4*
Dadul-papaja, 1:418
Daemonorops
  calapparius, 2:383*n6*, 3:639*n9*
  draco, 2:74, 2:318*n25*, 3:499*n3*,
    3:663*n4*, 5:540*n1*
  nigra, 3:643*n1*
  palembanicus, 3:653
  ruber, 3:666*n16*
Daemonum arbor, 2:570
Daisy, 4:249*n1*, 5:83*n3*
Dalbergia parviflora, 2:87*n21*,
  2:281*n3*, 2:314*n6*, 2:607*n4*,
  3:251*n22*, 3:499*n2*, 3:499*n4*,
  3:501*n19*, 3:511*n3*, 4:196*n9*
Dalundung, 2:349
Daluwan, 3:38
Dammar, 5:476
  Bangay, 2:269

batu, 2:259, 2:263, 2:264, 2:265–
  2:268, 2:269, 2:637, 5:469
Celebese, 2:269–2:270
Galla Galla, 2:242
Itam, 2:239
Leucomelaena, 2:260
Malejo, 2:257, 2:259
Puti, 2:266
Radja, 2:268
selan, 2:250, 2:253, 2:254–2:255,
  2:256–2:260, 2:261, 2:265,
  2:267, 2:295, 2:299, 3:667
Dammara
  alba, 2:263, 2:264, 2:265–2:268
  nigra, 2:239, 2:240–2:241, 2:242,
    2:243
  Selanica, 2:253, 2:254–2:255,
    2:256–2:260
Dangan-dangan, 1:417
Dangora, 3:299
Dansidan, 4:583
Danu caloway, 3:71
Daon tsju tsju rambu, 3:245, *3:246*,
  3:247
Dap Dap Tree, 2:350
Darnell, red, 5:27*n2*, 5:35*n3*
Dart root, 4:273, 5:299, *5:300*,
  5:301–5:305
Dart Tree, *2:396*
Dasyatis kuhlii, 5:71*n5*
Datura
  fastuosa, 2:541*n5*, 3:589*n8*,
    4:177*n8*, 4:182*n46*, 4:183*n3*,
    5:265*n4*
    alba, 4:177*n5*
  metel, 4:179*n17*, 4:180*n26*
  stramonium, 2:343*n2*
Daudau, 2:607
Dauläe, 3:308
Daulassi, 3:176
Daun
  api, *3:488*, 3:489–3:490
  apostema, 4:507
  assam, 2:509, 2:511, *4:524*,
    4:527–4:528
  besaar, 4:229

kecil, 4:229In3
   kitsjil, 4:229, *4:230*
bacong, 5:301
baggea, 1:290, 3:453, 3:455,
   3:456, 3:457
baru, 2:331, 2:332, 2:333, 2:343,
   2:387, 3:131, 3:133, 3:294*n10*,
   3:409, 4:551, 4:551*n5*, 4:553,
   5:52, 5:55
   mera, 2:337
benang, 5:95, 5:99
bidji katjang, 5:105
bissol, 4:479, 4:507, 5:65, 5:66
boaya, 3:373, *3:374*, 3:375, 3:468
boba, 4:553, 5:115
bonkus, *5:218*, 5:219–5:220
bosar, 3:286
bulan, 4:478
   babulu, *3:536*, 3:537
bulang, 3:533, *3:534*, 3:535
caju, 5:503
cambing, 3:153, 3:399, 4:255,
   5:80
candal, 2:536, 3:47, 3:65, 3:103
capialun, 4:515
capur, 3:131, 3:407, *3:408*, 31409
cardamom, 5:291
cassise, 4:297, *4:298*
casuari laut, 5:379
catsjo, 3:575
cattam cattam, 4:485, 4:487
caudal, 3:47
coijn, 3:301, *3:302*, 3:303
contu, 3:412, 4:138*n33*, *4:488*,
   4:489–4:490, 5:77
copo copo, 5:115
corpoo, 4:545, *4:546*, 4:563
   lacki lacki, 4:564
corra corra, 5:213, *5:214*, 5:215,
   5:219
cucuran, 4:234, 4:523, 4:529,
   4:530, 4:531
curap, 5:531
dammar angus, 5:71
djarak, 3:336
djarong, 5:71
dulang, 3:81
gatta gambir, 3:573, *3:574*,
   3:575, 3:582
   bulu bulu, 3:577–3:578

gattal
   ayer, 4:419
   babi, 5:90
   besaar, 5:89
   matti, 5:91
gindi, 3:661, 3:673–3:676
gora, 3:287
gora gora, 4:255
gorita, *3:348*, 3:349–3:350
gosso, 3:388, 3:405
grisek, 3:301
grissik, 3:305, *3:306–3:307*,
   3:308
hati hati, 4:251, 4:258
heran, 5:73
hode, 4:377
ila ila, 5:60
ingat, *5:52*, 5:55
jepit rambut, 3:247*n6*
kasturi, 3:243*n1*
kentut, 5:109*n18*
kiker, 3:402
kitschil, 5:80, 5:575
kitsjil, 3:153, 3:161, 3:286
knoto knoto, 5:77*n5*
kopo kopi, 5:115*n7*
labrang, 5:11, *5:12*, 5:13
lacat lacat, 5:114
lacca, 3:250, 3:499*n2*
lacum, 4:515
lada, 5:279
lauer, 5:255
layang layang, 5:150
lida lida, *3:466*, 3:467–3:469
lolâh munut, 3:468
maas, 3:286
manko, 3:301, *3:302*, 3:303
matta panas, 5:49, *5:50*, 5:51,
   5:52, 5:53, 5:577
medji, 3:81, 4:27
moreto, 5:101
mucca
   manis, 5:67
   sackit, 5:65
murit, 5:279
ngassi, 5:541
panu, 5:93
papeda, 3:301, *3:302*, 3:303
   pandjang, 3:305, *3:306–3:307*,
      3:308
   papua, 3:311

parawas, 3:142
petola, 5:175–5:176, 5:175*n*7
pittis, 4:549
 kitsjil, *4:550*, 4:551–4:552
 pittis, 4:503, *4:504, 4:505,*
  4:506
poding, *3:298*, 3:299–3:300
prada, *3:298*, 3:299–3:300, 4:163,
 5:63
 utan, 5:63
puppur, 4:255
putat, 3:100
putry, *3:370*, 3:371–3:372
rambu, 4:477, 5:157
russa, 5:69
sabran, 4:377, 4:377*n1*
sacandal, 3:47
sacko, 1:473
salam, 3:141, 3:141*n7*
salari bulang, 3:16
salawar, 5:111
saleh, 5:295
sanca, *3:394*, 3:395
sasappo, 5:86
sipat, 5:69, 5:83
sjurare, 5:60
sompittan, 5:301
subat, 5:189*n27*, 5:197, 5:211
 parampuan, 5:211
sudu sudu, 3:329
suggi, 3:315, *3:316*, 3:317
tali tali, 4:583
tassibeh, 4:79
tawas, 4:513*n16*
tickar, 3:443
tinta, 5:83
tocol manusia, 4:268
toteho, 3:47
tsjinta, 5:27, *5:78*, 5:79–5:81,
 5:80
ubi, 4:507
ulang ulang, 3:16
wawo, 5:255
Daussa, 3:293
 paia, 3:295
Däussa, 5:301
Dava karandi, 4:358
Davallia elata, 5:139*n2*

Davo piloc, 4:122
Dawan, 2:457, *2:458–2:459*, 2:460–
 2:461, 3:163
 Gabba Gabba, 2:460–2:461
 red, 2:457, 2:460
 stone, *2:456*, 2:457
Dawas Tree, 2:129
Dayngora, 3:287
De Bondt, Geeraerdt, 1:114
Debat guna, 5:67
Decaspermum fruticosum, 2:22*n5*,
 2:549*n2*, 3:161*n2*, *3:368n2*
Decus montium, 3:78
Deer plant, 5:69
Deeringia amaranthoides, 4:165*n1*
Dehaasia media, 2:532*n7*
Delima Laut, 2:582
Delima Tree, 2:137
Delphinium, 5:189*n16*
Demon's phallus, 5:253*n3*
Dendrobium, 5:212*n14*, 5:212*n17*,
 5:212*n18*
 acinaciforme, 5:189*n27*,
  5:197*n10*, 5:211*n9*
 anosmum, 5:201*n2*
 bicaudatum, 5:195*n1*, 5:197*n6*
 calceolum, 5:212*n20*
 confusum, 5:212*n18*
 crumenatum, 5:187*n7*
 ephemerum, 5:189*n14*, 5:197*n2*
 mirbellanum, 5:196*n14*
 moluccense, 5:211*n2*
 moschatum, 5:195*n6*
 papiloniferum, 5:201
 strebloceras, 5:195*n1*
 suaveolens, 5:189*n26*
Dendrocalamus
 asper, 3:195*n1*
 giganteus, 3:195*n5*
Dendrochirus zebra, 3:314*n6*,
 3:593*n3*
Dendrocnide, 2:81*n15*
 amplissima, 1:466*n5*
 longifolia, 1:466*n5*

Dendrophyllia
   gracilis, 5:439*n5*
   micranthus, 5:439*n*

Dentella repens, 4:534*n4*

Deringu laut, 5:353, 5:353*n5*

Derris
   elliptica, 3:347*n14,* 3:531*n1*
   heptaphylla, 3:585*n1*
   trifoliata, 3:539*n1*

Deryngo, 4:83

Desmodium
   gangeticum, 5:281*n10*
   triquetrum, 5:281*n12,* 5:282*n15*
   umbellatum, 2:206*n14,* 3:373*n1,*
    3:467*n6*

Despermum fruticosum, 3:161*n6*

Devil tree, 2:570, 3:27*n6*

Diadema setosum, 5:10*n16,* 5:379*n5*

Dialthaea, 3:225

Dialum indum, 2:135*n22*

Dianella odorata, 2:80*n10,* 2:88*n39,*
   2:109*n1,* 3:523*n6,* 5:9*n2*

Dianthus caryophyllus, 5:517*n3*

Dichapetalum moluccanum, 3:599*n6*

Dichocrocis punctiferalis, 3:338*n13*

Dicranopteris linearis, 2:131*n3*

Dictacophoron, 5:8

Dictyophora phalloidea, 3:177*n6,*
   3:527*n2,* 5:253*n1*

Digitaria
   adscendens, 5:211*n8*
   radicosa, 5:19*n3*
   sanguinalis, 5:6*n22,* 5:21*n1*

Dillem, 4:255

Dillenia, 2:505
   elliptica, 2:207*n1*
   serrata, 2:209*n7*

Dindinghay, 5:60, 5:61

Dinga dinga, 5:150

Dinochloa, 5:540*n1*

Dioclea
   hexandra, 3:479*n1*
   reflexa, 3:479*n1*

Dioscorea, 2:607*n5,* 4:506*n6*
   alata, 4:335*n1,* 4:335*n3,*
    4:335*n4,* 4:339*n1,* 4:343*n1,*
    4:345*n1,* 4:349*n7,* 4:351*n1,*
    4:441*n2,* 5:225*n19*
   bulbifera, 4:347*n1,* 4:347*n3,*
    4:359*n5,* 4:503*n5,* 4:506*n11,*
    4:507*n9*
   esculenta, 4:287*n10,* 4:353*n1,*
    4:479*n2,* 4:561*n4,* 5:213*n6,*
    5:225*n20,* 5:227*n4*
   hispida, 3:415*n7,* 4:290*n38,*
    4:359*n1,* 4:503*n2*
   nummularia, 4:494*n15,* 4:503*n3*
   pentaphylla, 4:357*n1*

Dioscurias, 4:221

Diospyros, 2:409*n1*
   ebenum, 2:427*n1*
   kaki, 1:398*n16*
   maritima, 2:415*n1*
   utilis, 2:415*n1*

Diplazium esculentum, 5:127*n1*

Diploknemia sebifera, 5:476*n8*

Diplophractum subobovata, 3:115*n1*

Dipper tree, 3:674

Dipterocarpus, 2:107*n18*

Disau Sayo, 3:109

Dischidia
   imbricata, 4:553*n1*
   nummularia, 4:551*n1*

Distichopora coccinea, 5:435*n3*

Divine plant, 5:171*n25*

Djagomutri, 4:105

Djahe, 4:51

Djahy, 4:51

Djalam garo, 3:247

Djali, 4:116

Djallo djallo, 5:315

Djamor bongkang, 5:233

Djangot, 5:457

Djarak, 3:336
   malacca, 3:341, *3:342,* 3:343

Djarong djarong, 3:357, *3:358,*
   3:359–3:360, *3:361*

Djati, 2:464

Djdurn dumung, 3:299

Djebis utan, 4:519

Djela djela, 5:25, *5:26*

Djendini, 2:631

Djerennang, 2:383

Djernang, 3:663, 3:664

Djerudju, 2:401, 5:311, 5:313

Djeruk, 2:207, 3:55, 3:65, 3:505,
    3:507, 3:508, 3:513, 3:514
    honey, 2:169
    lemon, 2:143, *2:144*, 2:145,
        2:146, 2:147–2:149, 2:151–
        2:152
    liemis, 2:155, *2:156*, 2:157,
        2:158, 2:159, *2:160*, 2:161,
        2:162
    Nipis, 2:209
    orange-apple, 2:163, *2:164*,
        2:165, *2:166*, 2:167–2:168,
        2:169, *2:170*, 2:171
    Purrut, 2:311
    water, 2:406
    wild, 2:153, *2:154*

Djingi, 4:443

Djintam
    Bengala, 4:219
    soa, 4:219

Djintan utan, 5:23, *5:24*

Djodjo, 3:211

Djolomadjico, *3:368*

Djomurut, 3:531

Dju malj, 3:371

Djudjambu, 2:305

Djudjaru, 1:313

Djuet, 3:142

Djujambu, 4:66

Docku, 1:287

Doctrine of similarity, 1:132

Dodeco, 1:333

Dodonaea viscosa, 3:161*n4*, *3:368n1*,
    3:393*n1*

Doffal madoro, 3:407

Dog ears, 5:49, *5:50*, 5:51, *5:52*,
    5:53

Dog Pepper, *3:548*, 3:549, 3:551

Dog's stones, 5:213*n2*

Dog's tongue, 5:65

Dogsbane, 4:541*n9*

Dokku, 1:242

Doleschallia polibete, 3:299*n1*

Dolichandrone spathacea, 2:199*n2*,
    2:541*n1*

Dolichos lablab, 3:103*n5*

Dolichus
    catjang, 4:401*n1*
    sinensis, 4:383, *4:384*, 4:385

Dombong lacki lacki, 5:108

Domestical Bune tree, 3:145,
    3:146, 3:147

Domu Hahocca, 3:203

Domul, 3:205

Donax canniformis, 3:215*n1*

Donderbaart, 5:195*n2*

Dongi dongi, 5:339

Donkey cucumbers, 5:367

Dooden-kruyd, 4:207

Dorghum nitidum, 5:35*n1*

Dormack Daging, 2:261–2:262

Dotaga, 5:115

Dotto Ducku, 3:471

Dou cerberi valli, 4:517

Dowara, 2:440

Draaken-bloet, 2:383

Dracaena angustifolia, 4:263*n5*

Dracontium, 4:295

Dracontomelon
    mangiferum, 1:429*n2*, 1:433,
        2:378*n4*, 2:474*n10*, 5:476*n3*,
        5:493*n2*
    sylvestre, 1:437*n1*, 2:205*n2*,
        2:541*n2*, 3:533*n2*, 4:445*n10*

Dracunculus, 3:347*n13*
    amboinicus, 4:298, 4:299–4:300
    aquaticus, 4:283, 4:299
    major, 4:304

Dragon tree, 1:433, *1:434*, 1:435, 2:*312*, 2:313, 2:384
  wild, *1:436*, 1:437–1:438
Dragon's Blood, 2:74, 2:318*n25*, 2:381, 2:*382*, 2:383–2:386, 3:663, 3:664, 3:666*n12*, 3:668–3:669
Driopteris, 2:222, 2:222*n29*
Drommel-boom, 3:27
Druis-gras, 5:*46*, 5:47–5:48
Drumstick Tree, 2:121, 2:*122*, 2:123–2:126
  wild, 2:127, 2:*128*, 2:129–2:130
Drymoglossum piloselloides, 5:205*n16*
Drymophloeus, 1:246*n34*
  appendiculatus, 1:285*n5*
  oliviformis, 1:285*n1*
Drynaria sparsiflora, 3:36*n15*, 3:382*n10*
Drynaria sparsisora, 1:271*n1*, 1:311*n4*, 4:581*n2*, 5:149*n1*, 5:189*n22*
Dryobalanops
  aromatica, 5:432*n18*, 5:585*n1*
  camphora, 5:585*n1*
Dryopteris, 5:139*n1*, 5:145, 5:157
  arborea, 5:*138*, 5:139 5:140
  campestris, 5:140
  ferox, 5:131*n1*
  petraea, 5:140
  sylvestris, 5:*138*, 5:139–5:140
  terrestris, 5:*138*, 5:139–5:140
Dsjerenang, 2:385
Duck plant, *4:226*, 4:227–4:228
Duckweed, 5:333
Dulci chinum, 5:11, 5:13
Dungur
Durein, 1:338
Duri carbou, 3:503, *3:504*
Durian tree, 1:335, *1:336*, 1:337–1:341
Durieon, 3:139
Durio, 1:338

Durio zibethinus, 1:337*n14*, 3:61*n2*, 3:139*n2*, 3:455*n4*, 4:323*n29*, 5:265*n2*, 5:473*n5*, 5:487*n3*
Durion
  babi, 1:337
  bornea, 1:337
  cassomba, 1:337
  manka, 1:337
Dutch Wood, 2:565
Dutra, 4:177, *4:178*, 4:179–4:182
  alba, 4:177
  nigra, 4:177
  rubra, 4:177
Duwe, 1:387
Dyer's plant, 5:95, 5:*96*, 5:97
Dysoxylum, 5:476*n1*
  amooroides, 5:476*n1*, 5:493*n1*
  euphleblum, 2:119*n1*
Dyudar, 2:570

E

Eagle Claws, 3:577
Eagle wood, 2:56, 2:57
Earth fruit, Japanese, *4:468*, 4:473–4:474
Earth's girdle, 5:*164*, 5:165
East Indian apple, 1:364, 1:367, 1:449
Eastern Wood Tree, 2:*518*, 2:519
Ebbenhout, 2:412
  bastaart, 2:428
  graauw, 2:415
  wilt, 2:419
Ebenum indicum, 2:411
Ebenus
  alba, 2:419, 2:428
  cinerea, 2:415
  femina, 2:415
  latifolia, 2:415
  molluca, 2:415
  parvifolia, 2:415
  silvestris prima, 2:419
Ebony, 3:271, 3:607
  bastard, 2:344, 2:346, 2:*426*, 2:427–2:429
  black, 2:*408*, 2:409–2:414

common, 2:*408*, 2:409–2:414
gray, 2:415, 2:*416*, 2:417
Ternatan, 2:409, 2:415, 2:*416*, 2:417
white, 2:409, 2:*418*, 2:419

Ecballium elaterium, 4:269*n17*, 4:445*n2*, 5:367*n24*

Echinochloa stagnina, 5:19*n8*

Echinometra
digitata, 5:422
mathaei, 5:445*n5*
setosa, 5:10

Echinus arboreus, 1:338

Eclipse plant, 5:*82*, 5:83

Eclipta prostrata, 5:69*n6*, 5:83*n1*

Ecliptica, 5:*82*, 5:83

Eel rope, 3:579, 3:*580*, 3:581–3:582, 3:*583*
broad-leaved, 3:582

Egg-Bearing tree, 3:121, 3:*122*, 3:123

Eggplant, 4:169*n1*, 4:171*n20*, 4:183*n3*, 5:259*n5*

Ekana, 3:81

Ekor
cutjing utang, 3:321, 3:*322*, 3:323
cutsing, 3:*318*, 3:319–3:320

Ela calli, 3:333, 5:582

Elaeocarpus, 2:522*n15*
crenatus, 3:67*n1*
glaber, 3:61*n1*
multiflorus, 3:67
oblongus, 3:61*n1*, 3:572*n36*
rumphii, 1:439*n3*, 3:67*n1*
sphaericus, 3:59*n1*

Elan, 4:351

Elateriospermum, 3:297*n1*

Elaterium, 5:367

Elder, 3:33, 3:34, 3:117, 3:153, 3:220, 3:309, 3:335, 3:341

Electrum, 5:551

Eleni, 1:196

Eleocharis dulcis, 5:11*n1*, 5:19*n4*

Elephant Tree, 2:351*n18*

Elephant's feet, 4:477*n7*

Elephants boom, 2:351

Elettaria
cardamomum, 4:43*n2*, 4:45*n8*
minor, 5:262*n15*

Eleusine
corocana, 5:18*n9*, 5:18*n11*
indica, 4:262*n17*, 4:538*n6*, 5:15*n13*, 5:17*n1*

Elfwort, 5:545*n4*

Elhanne, 3:250

Elytranthe amboinensis, 3:569*n9*

Elytrigia repens, 5:25*n4*

Ema wood, 2:*488*, 2:489

Emilia sonchifolia, 1:266*n8*, 4:261*n1*, 4:272*n53*, 5:67*n2*, 5:129*n4*

Empetrum
acetosum, 4:229, 4:*524*, 4:527–4:528, 5:364
rubrum, 4:138

Enckol, 4:45

Endospermum moluccanum, 2:387*n1*, 2:598*n8*, 3:364*n9*, 4:521*n6*

Enhalus acoroides, 5:353*n1*

Enkele Schoen-roosen, 3:221

Ensada tree, 2:603, 3:9, 3:25

Entada phaseoloides, 1:31, 2:442*n20*, 3:475*n1*, 3:476*n8*, 3:585*n3*, 3:631*n5*, 4:300*n11*, 4:475*n3*

Entu, 4:143

Eohur, 2:536

Ephemerum, 5:309

Epipactis, 5:179*n2*

Epipremnum pinnatum, 4:581*n1*

Epitonium scalare, 5:275*n1*

Equisetum, 5:379*n3*
aroborum squamatum, 5:172, 5:173–5:174
marinum, 5:*378*, 5:379
secundum, 5:173, 5:174
silvestre, 5:174

Eragrostis
  amabilis, 5:6*n22*
  tenella, 5:19*n5*

Eria
  moluccana, 5:201, 5:203*n15*
  monostachya, 5:206*n31*

Erica marina, 5:163, 5:381

Eriophoros javana, 1:502

Erva de figado, 4:262

Eryngium
  ceilanicum febrigugum, 5:313
  indicum, 5:311, *5:312*, 5:313–
    5:314
  maritimum, 5:311*n2*

Erysimum, 4:234*n10*
  cheiri, 3:269*n5*
  indicum, 4:234

Erythrina, *2:352*, *2:353*, *2:354*
  fusca, 2:355*n1*, 3:317*n8*
  variegata, 2:353*n1*, 3:417*n6*,
    3:479*n4*, 3:598*n5*, 3:626*n16*,
    4:168*n10*
  orientalis, 2:347*n1*

Essoyt Wood, 2:489 2:490

Esula esculenta, *5:100*, 5:105, *5:106*,
  5:129

Etrimelli, 4:45

Eucalyptus
  deglupta, 2:635*n1*
  sarassa, 2:635

Eucheuma, 5:170*n8*

Eugenia, 1:378, 2:21*n2*, 2:543*n1*,
  2:546*n10*, 2:548*n8*, 2:555*n1*,
  2:617*n6*, 3:95*n5*, 3:117*n3*,
  3:119*n2*, 3:141*n1*, 3:141*n2*,
  3:141*n4*, 3:401*n2*, 4:183*n2*,
  5:206*n28*, 5:207*n2*
  aquea, 2:543*n1*, 3:323*n9*
  caryophyllata, 2:3*n1*
  celebica, 1:379*n9*
  jambos, 1:379*n1*
  javanica, 2:189, 2:548*n9*, 5:468*n1*
  malaccensis, 1:333*n2*, 2:119*n4*,
    2:305*n20*, 2:513*n3*
  melastomifolia, 2:547*n1*,
    3:141*n11*

  rumphii, 2:3*n8*, 2:547*n4*, 4:183*n2*
  subglauca, 2:559*n1*

Eulophia, 5:223*n6*

Euodia
  amboinensis, 4:332*n7*
  latifolia, 2:281*n1*

Eupatorium, 3:247

Euphorbia, 3:329, 3:331
  antiquorum, 5:581*n2*
  atlantica, 3:333
  helioscopia, 5:105*n4*
  hirsuta, 5:581*n3*
  hirta, 5:105*n1*, 5:129*n8*
  lathyris, 2:359*n4*, 3:283*n4*
  longana, 1:432*n17*
  nerifolia, 2:189*n2*, 2:361*n12*,
    3:329*n1*
  tirucalli, 5:581*n1*
  tithymalus, 3:333

Euphyllia, 5:439*n4*

Eurycles amboinensis, 5:304*n24*,
  5:307*n1*

Eusideroxylon zwageri, 1:261*n27*

Evening flower, 3:225, 4:192

Ewasse Tree, 2:538, 2:539

Exocarpus epipjyllanthus, 5:503*n1*

Exoecaria agallocha, 2:88*n35*,
  2:359*n1*, 2:542*n7*, 2:584*n15*,
  2:639*n13*, 3:167*n2*, 3:332*n17*,
  3:411*n1*, 3:501*n23*, 3:502*n27*,
  5:202

Ey-keulen, 4:363

Eyes of the sea-crabs, 4:561, *4:562*

Eyes of the Starling, 3:567

Eyk, 5:483

Eyparehu tree, *3:66*, 3:67

Eytassi, 2:511

F

Faba
  aegyptia, 4:288, 5:327
    antiquorum, 5:321
  arborescens, 3:476
  coseira, 3:483
  cuseira, 4:420

de brhammane, 3:564
fresas, 4:406
marina, *3:474, 3:475–3:478,
   3:585, 3:637*
rubra, 4:399
Faenum graecum, 5:44
Fagara
   amboinensis, 4:565*n1*
   avicennae, 3:514*n17*
False Flax, 3:335
Fana, 4:25
Farnkraut, 5:129
Fart leaf, 4:489
Fartago lacki lacki, 5:18
Fat Tree, 2:376, 2:377–2:378
Fat-rope plant, 3:612
Fatahu Besaar, 2:367
Father of flies, *5:494, 5:497*
Fatta goma, 5:251
Faughio tree, 3:96, 3:96*n12*
Favidae, 5:437*n4*, 5:437*n5*
Favolus, 5:245*n6*
Favos dos paros sativo, 4:387
Favus saxeus, 5:437
Feast Flower, *3:217, 3:218, 3:219–
   3:221, 3:220n29*
Fenugreek, 1:485, 4:63*n41*,
   4:141*n5*, 4:233*n1*, 4:411*n2*,
   5:44*n17*
Feonum graecum, 4:233
Fern, 1:271*n1*, 1:311, 1:311*n2*,
   1:312, 1:313–1:314, *1:315*, 1:316,
   1:317, *1:318*, 1:319–1:320, 1:327,
   1:373*n17*, 3:24*n8*, 3:91*n3*
   bird's nest, 5:153*n2*
   edible, *5:127, 5:128*, 5:129–5:130
   fiddlehead, 5:125*n2*
   fire, 5:131*n3*
   flowering, *5:146, 5:147*
   maidenhair, *5:145n9*, 5:364*n14*
   male Ambonese, *5:128*, 5:131
   palm tree, *5:120, 5:121–5:123*
   rheumy, 5:136–5:137
   salamander, 5:151
   shrimp, 5:133*n2*, 5:136

water, *5:124, 5:125–5:126*
white, 5:133*n2*
wild, 5:122*n8*
writing, *5:159, 5:160, 5:161*
Feronia likonia, 2:198*n8*
Ferula galbanifera, 3:308
Festalis, 3:217*n3*
Fette fette, 5:301, 5:303, 5:305
Fever tree, 3:194
Fcvcrfcw, 5:71*n2*
Ficus, 2:334*n10*, 2:517*n5*, 2:611*n2*,
   3:3*n1*, 3:19*n1*, 3:37*n3*, 3:426*n9*,
   3:579*n5*, 5:105*n6*, 5:481*n2*
   adenosperma, 3:43*n2*
   altissima, 2:21*n4*, 2:406*n8*
   ampelos, 2:315*n16*, 2:374*n6*,
      2:398, 2:399*n9*, 3:490*n3*
   bengalensis, 3:24*n10*
   benjamina, 1:347*n3*, 2:43*n4*,
      2:47*n7*, 2:53*n6*, 2:65*n6*,
      2:222*n30*, 3:3*n2*, 3:8*n36*,
      3:23*n1*, 3:387*n2*, 3:494*n8*,
      3:511*n4*, 3:517*n2*, 3:571*n19*,
      3:571*n30*, 3:587*n2*, 3:615*n13*,
      4:175*n4*, 5:182*n48*
   botryocarpa, 3:43*n1*, 5:493*n4*
   calophylla, 3:11*n1*
   carica, 4:5*n16*
   corona, 3:43*n1*, 3:405*n6*
   drupacea, 3:4*n5*, 3:15*n2*, 3:405*n3*
   forstenii, 3:11*n4*
   haematocarpa, 3:24*n9*
   indica, 2:603, 3:4, 3:5, 3:7, 3:31,
      4:7
      veterum, 4:6, 4:6*n26*
   leucantatoma, 3:46*n13*
   malabaricas, 3:9
   melinocarpa, 3:405*n1*, 5:366*n19*
   moseleyana, 3:41*n4*
   nodosa, 3:33*n1*
   pilosa, 3:11*n4*
   punctata, 3:613*n1*, 3:613*n5*
   recurva, 3:607*n2*, 3:607*n5*,
      5:540*n1*
   religiosa, 3:27*n6*, 3:27*n7*
   rumphii, 3:27*n1*, 3:27*n6*
   septica, 2:595*n9*, 3:39*n3*, *3:44,
      3:45–3:46*, 3:45*n1*, 3:46*n13*,

3:118*n17*, 3:311*n6*, 3:581*n7*,
4:362*n15*, 5:15*n10*, 5:105*n7*,
5:245*n8*
sycomorus, 3:4*n4*
trematocarpa, 2:406*n10*, 3:375*n5*,
3:405*n5*
variegata, 3:34*n4*
vesiculosus, 5:346*n21*, 5:349*n5*
virgata, 3:15*n2*, 3:508*n21*,
4:563*n3*
wassa, 3:39*n1*, 3:118*n20*,
3:405*n4*, 5:493*n3*
Fidacca, 2:323
Fig tree, 1:344*n14*, 1:414, 1:419*n18*
Ambonese, *3:32*, 3:33–3:36
American, 3:9
East-Indian, 3:9
garden, 4:9*n4*
Indian, 3:7, 3:18, 3:331, 3:405,
4:5*n16*, 4:7*n45*
Malabar, 3:22, 3:23, 3:41, *3:42*,
3:405
noxious, *3:44*, 3:45–3:46
paper-giving, 3:37–3:38
rough, 3:39, *3:40*, 3:41
wild, *3:32*, 3:33–3:36, 3:39, *3:40*,
3:41
with green fuites, *3:42*, *3:43*
Figo
d'Orta, 4:9, 4:9*n4*
da India, 4:5*n16*
de Martabane, 4:6
Figuier d'Indi, 3:7
Filix, 1:102
amboinica mas sive agrestis, *5:128*,
5:131
aquatica, *5:124*, 5:125–5:126,
5:298
arborescens, 5:122
calamaria, 1:279, 2:131, 5:159,
5:160, 5:161, 5:221
canarina, 5:123, 5:150
esculenta, 3:100, 4:471
sive foemina, 5:127, *5:128*,
5:129–5:130
florida, *5:146*, 5:147
lanuginosa, 5:130
osmunda, *5:124*, 5:125–5:126,
5:297

repens, 5:159, *5:160*, 5:161
scandens perpulchra Brasiliana,
5:137
urens, 5:131
Fimbristylis, 5:4*n5*
setacea, 5:31*n1*
Fingernail Flower, 3:251*n16*
Finlaysonia obovata, 4:563*n1*
Fire root, *4:520*, 4:521–4:522
Fire skin, 5:483
Fire Tree, 2:105
Fish roe, *4:98*, 4:99–4:100
Fish-Slayer tree
beach, *3:166*, 3:167
mountain, 3:169, *3:170*, 3:171
Fisherman's Bark, *3:398*, 3:399
Fishhook rope, 3:573, *3:574*, 3:575
wooly, 3:577–3:578
Fissa Furu, 3:5
Flabellum
cinereum, 5:401
marinum, 5:374, 5:375–5:376,
5:378
cinereum, 5:407–5:408
planum, 5:375
multiplex, 5:375, 5:376
Flacourtica indica, 5:533*n1*
Flagellaria indica, 3:671*n1*
Flame of Fire, 3:299
Flame of the Forest, 3:357, *3:358*,
3:359–3:360, *3:361*
Flamma silvarum, 2:552, 3:357,
3:358, 3:359–3:360, *3:361*, 3:365,
4:522, 5:532
Flammulam, 5:334
Flannel plant, 5:513*n3*
Flat moon, 4:300
Flecke natoor, 4:511
Fles-appelen, 4:426
Flesh-Colored Resin Tree, 2:261–
2:262
Fleurya interrupta, 5:90*n5*
Flindersia amboinensis, 3:139*n1*

Flods manorae, 3:556, 3:557–3:561

Flora
  leydensis prodromus, 2:334*n13*
  malabar, 2:175
  malesiana, 3:3*n1*

Flores saxei, 5:436, 5:437–5:438

Flos
  cancri, 4:76, 4:79–4:80
  cardinalis, 3:371, 4:462, 4:465–
    4:467
  coeruleus, 3:562, 3:563–3:564,
    3:564
  convolutus, 3:324, 3:325–3:326
  cuspidum, 2:287, 2:288, 2:289–
    2:290
  festalis, 3:217, 3:218, 3:219–3:221
  flavus, 3:278, 3:279
  globolus, 4:246, 4:249–4:250
  horarius, 3:222, 3:223–3:224,
    3:230, 3:243, 4:192
  inpius, 4:247–4:248, 4:248
  manillanus, 3:327, 3:328
  mogorensis, 3:558
  perfulanus, 3:552, 3:553, 3:555
  principissae, 5:181
  siamicus, 5:570, 5:571
  Susannae, 4:243, 4:244, 4:245
  tanjonus, 2:287, 2:288, 2:289–
    2:290
  tigridis, 5:184
  triplicatus, 5:214, 5:215, 5:221–
    5:222

Floscopa scandens, 5:45*n21*

Flosculis verticillatis, 2:227

Focki focki, 4:169, 4:171

Foelha
  botaon, 4:250
  de inferno, 3:339
  de pavaon, 3:267
  de Sapatto Macho, 3:221
  mogari, 3:560
  o figo de inferno, 3:336
  sapato, 3:219*n16*
  usagre, 3:230

Foenum
  graecum, 3:281, 4:63, 4:141,
    4:411, 5:281
  marinum, 5:378, 5:379

Folhas de Trinitade, 2:351

Folho
  de contas, 4:79
  maco, 5:83
  margoseira, 4:449
  rabasso, 4:523

Folin, 4:499

Folium
  acidum, 2:507, 2:591
    majus, 2:509, 2:510, 2:511
    minus, 2:512, 2:513–2:514
  ardens, 2:387
  baggae, 1:219
  baggea
    maritimum, 3:455, 3:456,
      3:457
    verum, 3:453
  bracteatum, 3:285, 3:298,
    3:299–3:300, 3:393, 5:63, 5:95,
    5:95*n5*
  buccinatum, 4:29
    album, 4:29–4:30
    asperum, 4:30
    latifolium, 4:29
  calcosum, 1:279, 1:279*n33*, 3:81,
    3:407, 3:408, 3:409, 4:321
  causonis, 4:512, 4:515–4:517
  crocodili, 3:373, 3:374, 3:375,
    3:467
  foetoris, 4:489
  hircinum, 3:152, 3:153–3:154,
    3:399
  indum, 4:320
  intinctus, 3:141–3:142
  involutionis, 5:219
  linguae, 1:277, 3:466, 3:467–
    3:469, 3:579
  lunatum
    majus, 3:533, 3:534, 3:535
    minus, 3:527, 3:536, 3:537
  mappae, 3:80, 3:81
  mensarium, 4:30, 5:271, 5:273
    rubrum, 4:27
  petolatum, 5:172, 5:175–5:176,
    5:175*n7*
  pistorym, 3:451
  politorium, 3:404, 3:405–3:406,
    3:490, 5:366
  principissae, 3:370, 3:371–3:372

pruritus magnum, 5:89
repetitionis, 3:16
tinctorum, 5:95, *5:96,* 5:97
unicum, 5:295
urens, 3:174, 3:175–3:177
vivens, 4:268
Folkium polypi, 3:348, 3:349–3:350
Follis pinnatis, 2:227
Fongol, 3:79
Fool's herb, 4:123
Foolish herb, 4:125*n7*
Fougul, 3:73
Foula de Golfaon, 5:321
Four o'clock plant, 4:191*n1*
Foxtail, 5:377
Fragarius
    albus, 3:431
    niger, 3:427, 3:428, 3:429
    ruber, 2:485
Frangipani, 3:325*n1*
Frankincense, 2:59
Fraxinella, 2:198
Fraxinus bungeana, 3:96*n12*
Freycinetia
    funicularis, 3:461*n1,* 4:579*n7*
    gramineu, 5:39*n16*
Frog Shell, 3:311*n7*
Frog's house, 5:257
Fructus
    beretinus, 1:478
    bobae, 3:68, 3:69
    musculiformis, 2:278, 2:279–2:280
    regis, 5:525, 5:526, 5:527
Fruit of paradise, 4:6*n21*
Fruita
    balsa China, 5:119
    burruga, 4:506
        pennachoso, 4:351
    contsji, 3:568
    de Entrude, 3:49
    de gralha, 3:429
    gargantilla, 3:227
    Manilha, 2:437
    matta peyre, 3:528
    pemacho, 3:487

Frumentum, 4:115
Frutex
    aquosus, 1:275, 3:423
        femina, 2:511, *3:354,* 3:355–3:356
        mas, 3:351, *3:352,* 3:353
    carbonarius, 2:500
        asper, 3:403
        latifolius, 3:402
        primus albus, 3:400, 3:401–3:402
        primus ruber, 3:400, 3:401–3:402
    ceramicus, 3:396, 3:397
    cerasiformis, 3:418, 3:419
    excaecans, 3:410, 3:411–3:412
    indicus incognitus, 5:112
    linteanus, 3:376, 3:377–3:379
    marinus
        elegantissimus, 5:376
        ericae facie, 5:381
Fruticem globulorum, 3:625
Ful, 4:392
Fula
    arróz, 3:563
    aybana, 3:251
    betele, 2:296
    cananga, 2:296
    champacca, 2:303
    criqua, 3:563
    kauki tree, 2:289
    mogori, 3:558
    pavaon, 3:266
    sapato, 3:219
Fulha
    alacra, 5:181
    lacre, 5:181
    merenda, 4:192
    quadrohoras, 4:192
    rabicqui, 4:192
Fuluck, 3:189
Fungia
    fungites, 5:443*n1*
    scutaria, 5:443*n1*
Fungus
    arboreus, 5:245, 5:247
    arborum tuberosus, 5:251
    digitatus, 5:246, 5:247–5:248, 5:248

elati petasoide, 5:246, 5:247–5:248
elatus, 5:246, 5:247–5:248
   cochlearis, 5:246, 5:247–5:248,
     5:247*n1*
   igneus, 5:243, 5:249
   saxeus, 5:442, 5:443–5:444
     oblongus, 5:443, 5:444
   tuberosus, 5:251
Funis
   aquosus, 3:577, 3:582
   butonicus, 3:599
   convolutus, 3:584, 3:585–3:586,
     3:587
   cratium, 3:496, 3:497
   crepitans, 3:611, 4:507, 4:508,
     4:509, 4:510, 4:556, 4:561
     quartus, 4:510
     trifolius, 4:510
   dentarius, 3:605
   gnemoniformis, 3:484, 3:485
   incantatorius, 4:510
   muraenarum, 3:579, 3:580,
     3:581–3:582, 3:583
     latifolius, 3:578
   musarius, 3:134, 3:134*n12,* 3:602,
     3:603, 3:608, 3:637
   niger parvifolius, 3:599
   papius, 3:491, 3:492, 3:493,
     3:494–3:495
     parvifolius, 3:494
   perplexus, 3:471, 4:565
   polypoides, 4:556
   pulassarius, 3:524
   quadrangularis, 3:611
   quadrifidus, 3:472, 3:473, 3:485
   silvaticus, 3:477
   sirifolius, 3:553
   sylvaticus, 3:517
   toaccae, 4:565, 4:566
   uncatus, 3:573, 3:574, 3:575
     angustifolius, 3:575
     lanosus, 3:577–3:578
   urens, 3:488, 3:489–3:490
Funnel leaf, 4:523
Fuo, 4:405

Gabaä, 1:299
Gababango, 5:40, 5:43–5:45
Gabba gabba, 1:299, 2:315, 2:387,
   2:389, 3:113, 5:451
Gabi gabi, 3:72
Gada gada, 3:674
Gadhab, 1:197
Gadong, 4:359, 4:361
Gagufuru, 5:126
Gahnia rawacensis, 5:39*n19*
Gairo, 3:478
Gaja Mada, 1:36
Gajang, 1:455, 1:456, 1:457, 2:11
   laut, 2:85, 3:569
Gajati shrub, 3:280, 3:281–3:282
Gajatus, 1:455, 3:280, 3:281–3:282
Gajim, 1:455
Galanga, 3:215, 4:31, 4:32, 4:33–
   4:36, 4:54
   florida, 3:215
   major, 4:31, 4:32, 4:33, 4:35,
     5:261, 5:267
   Malaccensis, 4:76, 4:77–4:78
   mas, 5:271
   minor, 4:32, 4:33, 4:34, 4:35, 5:5
Galaxea fascicularis, 5:439*n2*
Galedupa Ceramensis, 2:85, 2:86,
   2:87–2:88
Galedupa Tree, 2:85, 2:86, 2:87–
   2:88
Galena, 3:313
Gali, 4:143
Galiassa, 4:35
Gall
   apple, 2:14
   forest rope, 3:609, 3:610, 3:611
Galli galli, 5:311
Gallinaria, 3:267
   acutifolia, 4:237, 4:238, 4:239
   rotundifolia, 4:237
Gallium, 3:335

G

Galuga Tree, 2:*114*, 2:115–2:117

Galussa de Levance, 3:528

Gambas, 4:511

Gambe, 4:133

Gambir, 3:491
 laut, 3:560, 3:617, 3:*618*, 3:619

Gambu puti, 1:379

Gamongan, 4:39

Gamuto, 1:399*n2*

Gancong utan, 4:*484*, 4:487

Ganda suli, 5:273

Gandapoura, 3:245, 3:247

Gandasulium, 4:*70*, 4:75
 red, 4:79

Gandola plant, 4:*456*, 4:457–4:460

Gangu, 4:461, 4:519

Ganiter tree, 3:*58*, 3:59–3:60
 with oblong fruit, 3:61, 3:*62*,
  3:63

Ganitrum, 3:*58*, 3:59–3:60
 oblongum, 3:61, 3:572

Ganja
 agrestis, 4:*128*, 4:131
 hortensis, 4:133, 5:69
 sativa, 4:*128*, 4:129 4:130
 utan, 4:*128*, 4:131

Ganoderma
 amboinense, 5:247*n1*
 cochlear, 5:242*n11*, 5:247*n1*,
  5:247*n3*

Ganse-distel, 4:261

Ganti java, 4:217

Garcinia, 1:391, 3:105*n1*
 amboinensis, 2:509*n1*, 3:350*n9*
 cambogia, 2:379*n6*, 4:57*n5*
 celebica, 1:392*n18*, 2:503*n3*,
  5:463*n2*
 ceramica, 2:513*n1*
 cornea, 1:394*n25*, 2:503*n1*
 dulcis, 1:394*n26*, 2:503*n9*
 hanguryi, 2:379*n6*
 indica, 4:140*n47*, 4:423*n10*
 latissima, 2:503*n7*
 mangostana, 1:389*n1*, 2:453*n16*,
  2:503*n2*
 picorhiza, 3:403*n3*

picrohiza, 1:276*n8*, 1:428*n7*,
  2:181*n15*, 2:201*n1*, 2:203*n4*,
  2:493*n3*, 2:509*n2*, 2:584*n12*
 rumphii, 1:392*n18*

Garden grass, 4:86

Gardenia augusta, 5:517*n1*

Garlic, 5:155, 5:220, 5:307, 5:307*n4*

Garlic Tree, 2:*118*, 2:119–2:120

Garlic, wild, 4:312

Garo, 2:*52*, 2:53–2:62
 Boaya, 2:363
 Capalla, 2:55
 Cominjan, 2:54, 2:56
 Eckor, 2:55
 Malacca, 2:*52*, 2:53
 Matta Buta, 2:363
 Ramas, 2:56
 Tenga-tenga
 Tingelam, 2:55
 Tsjampacca, 2:61

Garuga abilo, 2:205*n1*

Garundo pala, 3:528

Gastrochilus pandurata, 5:223*n11*

Gatsjo, 3:105

Gatta Gambir, 1:234

Gayang, 2:271, 3:57, 3:*58*
 Laut, 2:324

Geanthus, 5:266*n7*

Gedang castela, 1:417

Gedrummelte mutsen, 5:444

Geerst, 4:115, 4:115*n4*

Gelala, 3:479
 albe, 2:*352*, 2:353–2:354
 aquatica, 2:351, 2:355, 2:*356*,
  2:357
 Ayer, 2:355
 beach, 2:347, 2:*348*, 2:349–2:351
 black, 2:347, 2:*348*, 2:349–2:351
 flamed, 2:*352*, 2:353–2:354
 Itam, 2:349
 Laut, 2:349
 litorea, 2:347, 2:*348*, 2:349–2:351
 picta, 2:*352*, 2:353–2:354
 Puti, 2:353
 water, 2:355, 2:*356*, 2:357
 white, 2:*352*, 2:353–2:354, 4:168

Gelang
  babi, 4:215
  bengala, 5:317
  Chinese, 4:215*n2*
  earth, 4:215*n6*
  laut, 5:315, 5:317
  passayr, 5:317
  pig, 4:215*n4*
  puti, 4:215
  sand, 4:215*n8*
  tana, 4:215
  tsjina, 4:215
  white, 4:215*n2*

Gelidium amansii, 5:170*n8*

Gelzeminum arabicum, 3:558

Gember, 4:51

Genda
  pura, 3:241, *3:242*, 3:243–3:244,
    3:245
    agrestis, 3:244
  russa besaar, 2:285–2:286

Gendarussa, 2:333, 3:286, *3:292,
  3:293–3:294*, 3:569, 4:332*n7*,
  5:95, 5:99, 5:109, 5:471
  femina, 3:295, *3:296*, 5:282
  foeminam, 3:258
  Itam, 3:294
  nigra, 5:61
  parampuan, 5:99

Gengeli, 4:119

Gengibil, 4:51

Geniculatum, *5:16*, 5:17–5:18

Genista, 3:269*n3*, 5:283
  aculeata, 3:623, *3:624*, 3:625–
    3:626
  tinctoria, 4:231, 4:231*n1*, 5:283*n3*

Geno geno, 4:487

Gentiana, 5:221, 5:221*n3*

Gepalu, 3:345, 3:347

Gerennang, 2:383

Gering geringan, 4:232

Geroro mavalla, 3:427

Gevoel-kruid, 4:268

Geyduar, 4:39

Giba Tree, 2:285–2:286

Gigantochloa
  aspera, 3:195*n1*
  levis, 3:642*n24*
  verticillata, 3:191*n2*

Gigi Gadja, 2:627

Gigirang, 3:351

Gilded Leaf, *3:298*, 3:299–3:300

Gillyflower, 5:517
  stone, *5:436*, 5:437–5:438

Gindi zeytang, 3:674

GIndian Apple Tree, 2:172, 2:173–
  2:175

Gingay, 4:143, 4:144

Gingeli, 4:120

Ginger, 4:67, 4:182, 5:222*n11*,
  5:223, 5:313, 5:317, 5:346,
  5:430*n13*, 5:444, 5:511*n8*, 5:541,
  5:545. *See also Zingiber*
  big head, 4:69*n12*
  grassy, 4:55*n6*
  large, 4:49, *4:50*, 4:51–4:54
  large-rooted, 4:69
  rasp, 3:139
  rumphii, 3:139
  small, *4:50*, 4:55–4:56
  war, 4:51
  West-Indian, 4:54
  wild, 4:39, 4:40, *4:44*, 4:48
  yellow, 4:60

Gingilena, 4:119

Ginji, 4:123, *4:124*, 4:125–4:127,
  4:180
  Papua, 4:125, 4:126

Ginseng, 5:545, *5:546*, 5:547–5:553

Girasole, 3:336

Girdle of stones, 4:528*n9*

Girgilim, 4:122

Girgir, 4:392

Girl's Leaf, 3:371, 3:389

Giroffeli, 2:5

Girusal, 1:348

Gitta
  gambir, 3:575
  jemu, 2:379

Gladiolus
  indicus, 4:79
  odoratus, 2:80, 4:89, *4:90*, 4:91,
    5:9
    indicus, 2:88*n39*, 4:91
Glans terrestris costensis, *4:376*,
  4:377
Glass plant, 5:47*n3*
Glasswort, 5:315*n7*
Glaucium dioscoridis, 4:455
Gleichenia linearis, 1:279*n30*,
  2:105*n7*, 5:159*n1*, 5:221*n10*
Globba, 5:219
  acris, 5:269
  baccifera, 1:279
  boppolulo, 5:266
  crispa, 5:261, 5:263
    rubra, 5:260, 5:264, 5:265,
      5:266
    viridis, 5:260, *5:264*, 5:265–
      5:266
  durian, 5:263, 5:267, 5:274
    mera, 5:219*n6*, 5:273
  durium idju, 5:265
  hatuana, 5:266
  longa, 5:258, *5:260*, 5:261–5:263
  maccassarensis, 5:261
  maraja, 5:273
  marantina, 4:41*n1*, 4:365*n4*
  papua mera, 5:263
  repens, 5:273
  silvestris
    florida, 5:273
    major, 5:270, 5:271, 5:272,
      5:273–5:274
    minor, 5:273
  subterranea, 5:273
  utan, 5:266, 5:271, 5:273
    besaar, 5:271
  uviformis, *5:258*, 5:261, 5:264,
    5:267–5:268, 5:273, 5:274
Globularia, 2:43, 2:45, 2:46
Globuli majores, 3:627, *3:628*,
  3:629–3:630
Glompan Tita, 3:78
Gloriosa superba, 5:310*n8*
Gluta ranghas, 2:391*n1*

Glyceria fluitans, 5:337*n7*
Glycine max, 4:409*n1*
Glycyrrhiza, 3:567, 5:550*n43*
  glabra, 3:564*n21*
Gmelina villosa, 2:183*n1*, 2:187*n1*,
  5:468*n2*, 5:539*n1*, 5:539*n4*
Gnemon
  tali, *3:486*, 3:487
  utan, 1:481
Gnemon, 1:244
  domestica, 1:477
  funicularis, *3:486*, 3:487
  silvestris, 1:481
Gnemon shrub, 3:485
  rope-like, *3:486*, 3:487
Gnemon tree
  domestic, 1:475, *1:476*, 1:477–
    1:478, *1:479*
  forest rope, 1:481
  vine, 1:481
  wild, 1:481–1:482
Gnetum
  gnemon, 1:244*n31*, 1:475*n1*,
    1:481*n1*, 2:205*n4*, 2:377*n2*,
    2:509*n4*, 3:61*n6*, 3:431*n5*,
    3:485*n1*, 3:485*n2*, 4:362*n17*,
    4:582*n7*, 5:491*n2*
  indicum, 3:487*n1*
Goa, 1:226
  tsjay, 4:261
Goat Leaf tree, *3:152*, 3:153–3:154
Goat plant, 5:87*n11*
Goat Tree, 2:205–2:206
Gobbu, *5:40*, 5:43–5:45
Gocarni, 3:564
God's Tree, 2:635
Goddam, 5:18
Godless flower, *4:246*, 4:247–4:248
Gofasa batu, 2:451*n3*, 2:451*n8*
Goffo
  hayran, 5:73
  heka, 5:49, *5:50*, 5:51, *5:52*, 5:53
  miskyn, 5:80
  soninga, *5:52*, 5:55
  suba

Gofforu, 3:24

Goheka mafalla, 5:257

Gohi, 3:41, 3:43
  daun bezaar, *3:40*, 3:41

Gohungung, 3:621

Golden Shower, 2:121*n1*

Gomassi, 1:356

Gomo, 1:352

Gomphrena globosa, 4:249*n3*

Gomuho, 5:141

Gomuto tree, 1:241, 1:271–1:273,
  1:274, 1:279, 1:280

Gondo, 3:84, 3:476, 3:629

Gondola
  alba, 4:456, 4:457–4:460
  rubra, 4:456, 4:457–4:460

Gongay, 3:635

Goniothalamus, 2:299*n1*

Gonystylus bancanus, 2:88*n34*

Gora, 1:372
  jadi, 1:377

Goraka, 4:51

Goram, 1:463

Gordonia rumphii, 3:143*n1*

Gori-maiho, 5:170

Gorogoro, 2:370

Gorse, 5:377

Gorumi dodo, 5:60

Gossali, 2:546

Gossipium, 3:234
  acuminatum, 3:239*n1*
  brasilense, 3:239*n1*
  daemonis, 3:240, 3:241
  indicum, 3:233*n1*
  latifolium, 3:238, 3:239
  purpurascens, 3:233*n1*, 3:234*n3*

Gossongi, 5:353

Gote kelingan, 5:11, *5:12,* 5:13

Gotnemsegiar, *3:238*

Gotte kelingan, 4:377

Goudal, 3:34, 3:36

Goudong Tree, 2:350, 2:351

Gouggong, 3:621

Gourd, 4:421, *4:422,* 4:423, *4:424,*
  4:425–4:426
  bottle, 4:354*n2*, 4:425*n2*
  dishcloth, 4:437*n1*
  Indian, 4:427, *4:428,* 4:429
  wax, 4:421*n1*

Gowa, 1:37

Grabosa
  ovada, 4:363
  rodonda, 4:358

Gracilaria
  edulis, 5:170*n8,* 5:339*n3*
  lichenoides, 5:340*n12,* 5:341*n1,*
    5:358*n13,* 5:433*n21*

Gragarius ruber, 3:424, 3:425–3:426

Grain de Molucca, 3:345, 3:346

Grain de molucco, 5:573

Grain de Musce, 3:244

Gramen
  aciculatum, 5:23, 5:24, 5:45,
    5:129
  anatum, 5:22
  arguens, 5:26, 5:27
  caninum, 5:19
  capitatum, 5:13, 5:15
  caricosum, 5:32, 5:33–5:34, 5:37
  cruciatum, 5:18
  cyperoides, 5:4, 5:5
  equinum, 5:25, 5:26
  fumi, 5:19
  paronychiale, 5:13, 5:15
  repens, 5:18
    minus, 5:19–5:20
  roris, 5:21–5:22
  supinum, 5:16, 5:17–5:18
  supplex
    5:21–5:22
  tremulum, 5:23, 5:24
  vaccinum, 5:16, 5:17–5:18
  venti, 5:19

Gramen mannae, 4:116

Grammatophyllum, 5:180*n8*
  scriptum, 1:328*n38,* 2:222*n27,*
    4:563*n5,* 5:175*n7,* 5:179*n5*

Grana tiglia, 3:347

Granaat-Appel, 2:137

Granatum litoreum, 2:580, 2:581–2:582, 2:583, 2:584–2:585

Granum
  indicum, 3:564
  moluccum, 3:344, 3:345–3:347, 3:347
  moschatum, 2:527, *3:242,* 3:243–3:244, *3:244,* 4:63
  sanel, 2:527

Grape tree, 1:429, 2:301, *2:302, 2:303–2:305*

Grape vine, 3:47

Grapes, monkey, 4:519*n3*

Grapevine, wild, *4:518,* 4:519

Graptophyllum pictum, 3:299*n1,* 3:355*n4,* 3:389*n3,* 3:393*n1,* 4:168*n12,* 4:507*n7,* 5:63*n2,* 5:95*n5*

Grass
  Bermuda, 5:19*n1*
  beseeching, 5:21–5:22
  cow, *5:16,* 5:17–5:18, 5:19
  crab, 5:19*n3*
  creeping, water, 5:19
  cutting, 5:33*n9*
  dew, 5:21–5:22, 5:21*n5*
  dog, 5:19
  eyelash, 5:34*n15*
  fisheye, 5:15*n4*
  foxtail, 5:34
  headed, *5:13,* 5:15
  horse's, 5:23, *5:24,* 5:25, *5:26*
  katjang, 5:105
  manna, 5:337*n7*
  mossy, 5:31, i5:32
  pin, 5:23, *5:24*
  razor, *5:32,* 5:33–5:34, *5:36,* 5:37–5:39
  reed, *5:26,* 5:29–5:30
  reproach, *5:26,* 5:27
  rye, 5:27*n2*
  sea, 5:337*n3*
  serah, 5:35*n2*
  small, creeping, 5:19–5:20
  smoke, *5:16,* 5:20
  tillage, 5:25, *5:26*
  water buffalo, 5:17*n2*
  Whitlow, 5:15*n5*

wind, 5:19, 5:19*n7*
wire, 5:19*n1*

Grasshopper tree, *3:158,* 3:159

Gratiolae affinis Americana, 5:87

Gratiosa, 5:27

Great Palm, 1:201*n3*

Great wild forest rope, *3:606,* 3:607–3:608

Greens
  bamboo, 5:43*n4*
  coconut, 5:298*n7*
  coral, 5:339*n6*
  goat, 5:103
  rogues, 5:44
  royal, 5:155
  shrimp, *5:68,* 5:69
  sweet, 5:298*n6*
  water, 5:335, *5:336*
  whore's, 5:44

Greenweed, 5:283*n3*
  dyer's, 4:231*n1*

Gressek, 3:16

Grewia salutaris, 3:79*n16*

Griza, 4:107–4:112

Groene katjang, 4:405

Groot
  hertstong, 3:382–3:383
  maagdekruyd, 3:483
  plompe, 5:321

Groote
  praatjes, *3:627, 3:628, 3:629–3:630*
  taan-boom, 3:119

Grossularia
  alba, 3:19
  domestica, 3:14, 3:15–3:16, *3:17,* 3:18, 4:563
  silvestris, 3:19
  wild, 3:19, 3:20, 3:21

Grote Peterzely-boom, 3:311

Growing nest, 5:229, *5:230,* 5:231

Guada, 3:206

Guaiacum officinale, 2:123*n4,* 2:431*n9,* 4:495*n19,* 4:499*n16*

Guajave tree
　domestic, *1:402*, 1:403–1:405
　forest, *1:406*, 1:407–1:409
　small, *1:410*, 1:411
Guanabana, 1:397, 1:397*n12*
Guanamba, 3:244
Guapariba, 2:603
Guatteria rumphii, 2:421*n1*
Guayavo, 1:404
Guettarda speciosa, 2:473*n1*
Gujavos, 3:141, 3:145, 3:159, 3:255
Gukacktsjil, 5:539
Gula, 1:266
Gulong Tsjutsju Tree, 2:191
Guluga, 3:314
Gum-Apple Tree, 2:326
Gumi
　benjamin, 3:559*n15*
　momadi, 3:542
　rotto rotto, 4:507
　susu, 3:542
　uccu, 3:490
Gumira tree, *3:152*, 3:153–3:154,
　3:595
　beach, 3:155, *3:156*
　laut, 3:155, *3:156*
Gumma Sari Sari, 2:497
Gummi
　Ammoniacum, 2:380
　Animae, 2:269
　cerasorum, 3:476
　Elemi, 2:227, 2:237
　Gutta Gumba, 2:379
　lacca, 3:250, 3:251, 3:501, 4:196
Gumoha gogila, 5:141
Gunpowder tree, 5:514*n9*
Gura bata, 3:301, 3:311
Gurama cussu, 4:86
Gurken, 4:432
Gutta Cambodje, 2:365, 2:379–
　2:380, 2:380
Guvajos shrub, 3:615
Gymnacranthera zippeliana, 2:43*n6*

Gymnanthes indicum, 3:167*n1*
Gymnema syringaefolium, 3:597*n1*,
　4:543*n3*
Gymnopus microcarpus, 5:242*n12*,
　5:249*n2*
Gynandropsis pentaphylla, 4:233*n4*,
　5:109*n21*
Gynja, 4:203
Gyrinopsis brachyantha, 5:489*n1*

H

Haät, 3:34
　Adam, 3:41
Haback, 4:207
Habaco-nabati, 4:207
Habel
　assis, 5:13
　zelin, 5:13
Habenaria rumphii, 4:245*n12*,
　5:227*n1*
Hacca leylu, 5:49, *5:50*, 5:51, *5:52*,
　5:53
Hachal Indi, 4:192, 4:193
Haedera, 3:607
Haematoxylum
　brasiletto, 3:272*n36*, 3:501*n24*
　campechianum, 3:272*n36*
Haffu
　Boppolulo, 2:250
　Gogira, 2:250
　Haffu Xula, 2:250
Hahay Muli Tree, 2:127
Hahumalossey, 2:81, 2:234
Hahuol, 3:41, 3:43
Hahuru, 5:122
　puti, 5:122
　　riri ulen, 5:123
　riri ulan, 5:122
　yar, 5:150
　yar lari, 5:123
Hair leaf, 4:477
Hair of the nymphs, 5:337–5:338
Hairpin Plant, 3:245, *3:246*, 3:247
Hajam, 1:455

Halale, 4:507

Halat, 3:401

Halau sussuela, 5:126

Halauila, 5:126

Halaur Tree, *2:534*, 2:535–2:536, *2:537*

Halaut Kitsjil, 2:536, *2:537*

Halcyonium
   album, 5:449
   rubrum indicum, 5:420, *5:423–5:424*

Halecus
   litorea, *3:130*, 3:131, 3:407
   terestris, *3:132*, *3:133–3:134*

Haleky tree
   beach, *3:130*, *3:131*
   coarse, 3:133
   daun bezaar, 3:134
   land, *3:132*, *3:133–3:134*
   lau muri, 3:134
   laut, 3:131
   puti, 3:134
   terrestris, 3:321
   white, *3:133*

Halicacabus, *5:106*, *5:115*, *5:116*, *5:117*, *5:119*
   baccifer, *5:116*, *5:119*
   indicus, 4:513, *5:115*, *5:116*, *5:117*
   peregrinus, *5:115*

Halomitra pileus, 5:443*n5*

Hamuki, 5:559

Hamzâ's Tree, 2:17

Hanet Tree, 2:311*n16*, 2:501–2:502

Hange hange, 5:153

Hanile Oöa, 3:78

Hanoa, 3:118

Hapughaha Tree, 2:305

Hareoman, 4:103

Harienhaur uun, 5:340

Harlaar, 3:487

Harpullia arborea, 2:446*n5*

Hartstongue, 5:364*n5*

Haru
   Cou, 2:337
   Layn, 2:340
   Paracca, Tali2:335

Hary raya, 3:219

Hascish ghesale, 4:87

Hassat, 3:490

Hatabessi, 2:446

Hataul Lau-muri, 2:329

Hataur, 2:323

Hate
   Bessi, 2:431
   bula, 1:495
   fica fica ngadu, 3:407
   gou, 3:373
   java, 3:27
   jujo, 3:101
   Kelang, 2:106
   papeda, 3:351
   Sasoae, 2:549
   sinangi, 3:423

Hatta, 5:141

Hattou-Ui, 2:323

Hatu, 1:460
   callan, 4:528
   ehu, 4:528
   hanal, 4:528
   luli, 5:414

Hatuëhu
   cau, 5:421
   pute, 5:414

Hau ettut, 4:489

Haul Amuret, 3:72

Haulawan Tree, 2:283

Haun
   laynulun, 3:455
   pantey, 3:455, *3:456*, *3:457*
   wassi, 3:453

Hauthy, 5:337, 5:337*n6*

Hawkweed, mouse-ear, 5:285*n2*

Hawthorn Tree, Indian, 2:189

Hawthorn, Batavian, 5:569

Haxis cachule, 4:87

Hay tsjay, 5:170, 5:171

Haylale, 4:373, 4:475, 4:477, 4:478
  biru, 4:483
Haytsjiu, 5:359
Hayuro, 4:361
Hazelwort, 4:523*n4*
Healing Bark, 3:517, *3:518,* 3:519–
  3:520
Heart-Fruit Tree, 2:*190,* 2:191–
  2:194
Heavenly dew grass, 4:116
Hebenaster, 2:*426,* 2:*427*–2:429
Heckeria umbellata, 5:259*n1*
Hedera, 3:607*n1*
  helix, 4:523*n3*
Hedychium chrysoleucum, 4:75*n1,*
  5:273*n18*
Hedyotis, 5:47*n3*
  tenelliflora, 5:47*n2*
  verticillata, 5:47*n1*
Hedysarum, 3:374, 3:375
Heers, 4:115, 4:115*n4*
Heilbonen, 4:403
Helenium indicum, 5:*106,* 5:107–
  5:109, 5:*108*
Heli
  mackilaun, 4:339
  tallan, 4:506
Helianthemum, 5:87
Helicia serrata, 5:499*n1*
Heliconia bihai, 1:356*n15,* 3:81*n5,*
  4:19*n5,* 4:27*n2,* 4:27*n4,* 4:*28,*
  4:29*n8,* 5:271*n5*
Helicteres isora, 5:525*n1*
Helinja, 4:343
Heliotrope, 3:*384,* 3:385
Helleborine
  molucca, 5:*178,* 5:179–5:185
  recentiorum, 5:195
Helleborus, 5:299*n4*
  albus, 4:59*n7,* 4:61, 5:201, 5:221,
    5:299
  amboinicus, 5:214, 5:221–5:222
  mas, 4:247
Hellula
  undularis, 4:235*n6*

Helminthostachys zeylanica,
  3:350*n11,* 4:367*n14,* 5:297*n1*
Hemels
  stut, 5:295
  wyzer, 5:295
Hemiepiphytes, 3:3*n1*
Hemigraphis, 3:300*n6,* 5:57*n2,*
  5:59*n3,* 5:59*n4,* 5:95*n7,* 5:129*n7*
  angustifolia, 3:55*n8,* 4:262*n8,*
    5:57*n3*
  petola, 5:58*n7*
  reptans, 5:53*n23,* 5:59*n1,* 5:99*n9*
Hemionitis, 5:141, 5:142, 5:143
Hena hena, 1:239
Henbane, 4:117*n3,* 4:171*n15,*
  4:180*n26*
Henna, 1:455*n2,* 2:631, 3:250,
  3:254
Herb of the Dead, 4:207*n12,* 4:211,
  4:212
Herba
  admirationis, 5:73, 5:74, 5:75
  amoris, 4:268
  colubraria, 5:105
  discipulorum, 5:279
  filorum, 5:56, 5:57–5:58
  lorotia, 4:87
  mascadensem, 4:86
  memoriae, 5:52, 5:55
  mimosa, 4:270
  moeroris, 5:78, 5:79–5:80
  moluccanam, 4:470
  necessitatis, 4:262
  papilionum, 5:65
  paronychiae, 4:534
  sentida, 4:268
  sentiens, 4:264, 4:267–4:272,
    4:271, 4:272, 5:79, 5:80, 5:129,
    5:283
  spiralis, 5:275, 5:277, 5:554,
    5:559
    hirsute, 5:559*n1*
    laevis, 5:559*n1*
  stultorum, 4:125
  supplex
    femina, 5:211
    major
      prima, 5:211
      quarta, 5:212

quinta, 5:212
  secunda, 5:212
  tertia, 5:212
  minor, 5:208, 5:210, 5:211–
    5:212
  timoris, 4:535
  verecunda, 4:271
  vitiliginum, 5:92, 5:93–5:94
  viva, 4:268, 4:270, 4:271
  zelotypiae, 5:111
Herba crinalium, 3:245, 3:246,
  3:247
Herba D'Cobres, 5:105
Herba do rosario, 5:41
Herba sentiens, 3:266
Herbam
  polypi, 5:291
  sacram, 5:53
Herbe de Musch, 3:244
Heri
  assapin, 4:339
  malona, 4:336
  manu, 4:336
  soun laun, 4:339
Heritiera littoralis, 1:460n8,
  1:465n3, 2:587n1, 2:602n18
Hernandia, 2:390
  ovigera, 3:36n14, 3:121n2,
    3:121n3
Herpetica, 5:530, 5:531–5:532
Heterocentrotus mammillatus,
  5:422n11
Heterogorgia magna, 5:375n1
Hexamitodera semivelutina, 2:12n15
Hiber Tree, 2:100, 2:101–2:102
Hibiscus, 1:432, 3:217n1, 3:217n2,
  3:218, 3:222, 4:510n14, 5:517
  mutabilis, 3:223n1, 4:192n12
  rosa-sinensis, 1:505n4, 3:217n4,
    3:294n6, 3:300n6, 3:559n10,
    4:138n37, 4:263n7, 4:460n12,
    4:521n2, 5:53n20, 5:67n5,
    5:126n15, 5:253n5, 5:279n3
  surattensis, 3:245n1, 3:245n3
  tiliaceus, 2:331n1, 2:333n7,
    2:335n1, 2:387n2, 3:91n2,

  3:131n2, 3:220n22, 3:224n8,
    3:233n2, 3:294n10, 3:409n10,
    4:551n5, 4:584n17
Hieracium, 5:285n1
  pilosella, 5:285n2
Hila, 1:62–1:63
  abbal, 4:281
  niä, 4:282
Himanthalia elongata, 5:346n22
Hina paretti, 3:224n9
Hinan, 3:118
Hippogrostis Amboinica, 5:23, 5:24,
  5:25, 5:26
Hippuris vulgaris, 5:414n4
Hirneola auricula judae, 5:243n16
Hissa, 4:86
  uren, 3:403
Hitu, 1:36, 1:37, 1:41, 1:44–1:46,
  1:327
Hoamoal, 1:46–1:48, 1:53, 1:54,
  1:55
Hoautli, 4:160
Hobnium, 2:412
Hodueg, 5:6
Hoedanig, 3:588, 3:589 3:591
Hohotu aman, 5:16, 5:17–5:18
Holly
  crab, 5:311n7
  Eastindian, 5:311, 5:312, 5:313–
    5:314
  sea, 5:311n2
Hollyhock, 3:217n2, 4:127n32
Holo Holo, 3:121
Holus marinum, 5:170n11
Homalanthus populneus, 3:411n2
Homalium foetidum, 2:443n1,
  2:452n14
Homalomena
  aromatica, 4:299n6
  cordata, 4:283n2, 4:299n1,
    4:299n6, 4:555n6, 5:335n7
Hondst-votten, 1:446
Honeysuckle, 3:220n30

Hooti, 5:372

Hopeck, 5:235

Hops, 4:449*n9*

Horehound, 5:73*n3*
  black, 4:255*n3*

Hornstedtia
  minor, 5:262*n15*
  rumphii, 5:261*n1*

Hornwood, 2:*504,* 2:*505*

Horse elder, 5:545*n4*

Horse Wood, 2:199*n2,* 2:*540,*
  2:*541–*2:*542*

Horse's hoof, 4:*523,* 4:*524,* 4:*525*

Horse-radish tree, 1:483*n1*

Horsetail, 5:379*n3,* 5:405*n2,* 5:411
  Ambonese, 5:*172,* 5:*173–*5:*174*

Horsfieldia, 2:43*n1,* 3:71*n4*
  canariformis, 2:43*n6*
  sylvestris, 2:39*n1,* 3:599*n2*

Hortus Malabar, 2:175

Hotton, 4:115

Hottong
  bengala, 4:116

Houboubo, 3:184

Houseleek, 4:144*n14,* 4:221*n3,*
  4:227*n1,* 5:195, 5:195*n2,* 5:205

Hoy-Gjek, 2:*383,* 2:*384*

Hoya
  alba, 4:547*n4*
  elegans, 4:549*n6*
  lutea, 4:541*n5*
  rumphii, 2:373*n3,* 3:497*n2,*
    3:571*n18,* 4:547*n1,* 5:205*n18,*
    5:513*n5*
  sussuela, 4:539*n1*

Hsiang-chiao, 4:6*n39*

Hua
  alan, 1:244
  alang, 1:239
  ewan, 1:239, 1:246
  keker, 1:242, 1:246
  niwel, 1:239
  soil, 1:244
  tette, 1:244
    ewan, 1:244

Huae, 3:391

Huan tayn, 3:443

Huca paretti, 3:224

Huda, 1:299

Hudu ambado, 1:441

Hudura pullu, 5:*23,* 5:*24*

Hulang, 5:*32,* 5:*35*

Hulen, 4:499

Hulong, 1:277, 5:215

Hulstkrabben, 5:311

Humelen, 2:561
  Malona, 2:*563,* 2:*564*

Hun, 4:150

Hunut, 3:113, 3:115

Hureck, 1:439

Hurong Tree, 2:111, 2:*112,* 2:113

Hussur, 1:394*n25,* 2:499, 2:503,
  2:*504,* 2:*505*
  Loumahum, 2:503

Hutta
  ahuae, 5:105
  ayan, 5:19–5:20
  dome, 5:39
  gindam, 5:31, i5:32
  laussa, 5:71
  tomar, 5:21–5:22
  urutta, 5:113
  wonin, 5:159

Huttum, 3:96
  darat, 3:100
  leymuri, 3:100

Hyaa, 4:203

Hyacinth, 3:313, 4:241, 5:307*n4*

Hyacinthaceae, 5:307*n2*

Hyalonema, 5:457*n1*

Hydnophytum amboinense, 5:229*n2*

Hydrocotyle asiatica, 4:523*n1*

Hyhetu Eër, 3:151

Hymenomycetineae, 5:242*n11*

Hyoscyamus, 4:117*n3,* 4:179*n17*

Hypericum, 3:627*n1*

Hyphaene coriacea, 2:76

Hypnea divaricata, 5:346n14

Hypolytrum latifolium, 5:38n15

Hypophyta, 5:261

Hyppopjobas, 5:551

Hyppuris saxea, 5:414

Hyris, 5:179

Hyssopus, 4:199n6, 4:205n4

Hystrix frutex, 5:510, 5:511

I

Ianthella basta, 5:451n1

Ibimani, 4:474

Ibira pitanga, 3:272

Ibu cuning, 4:57, 4:61

Ican sambilang, 5:539

Ichthyoctonos
    litorea, 3:166, 3:167
    montana, 3:169, 3:170, 3:171

Ide ide malacco, 3:567

Ido wahu, 4:99

Iger murri, 5:527

Ignix silvarum, 2:390

Igo igo, 4:268, 4:272

Ihrid, 4:136

Ihur tree, 1.268, 1.269

Ikan Swangi, 3:314n6

Ila â un, 3:405

Ilat boaya, 4:221

Ilex aquifolium, 5:311n1

Ily mullu, *5:8*

Ilysanthes antipoda, 4:533n1

Ima castela, 4:370

Imgi, 4:51

Imie, 3:250

Immortal plant, 5:171n25

Impatiens balsamina, 3:251n20,
    3:254n40, 4:195n2, 4:197n14,
    4:269n18

Imperata cylindrica, 1:503n10,
    3:463n2, 4:54n13, 4:584n9,
    5:5n18, 5:33n1, 5:37n6,
    5:223n18, 5:497n6

Ina
    alang, 4:285
    appar, 4:287
    kekermoni, 4:287
    kekin mahina, 4:287
    massahöut, 4:287
    tohalo, 4:287

Inaja Guacuba, 1:199, 1:208

Inan, 4:133

Indang, 3:435

Indian bat tree, 3:633

Indian Bean, 3:483

Indian Job's tears, *4:98*, 4:100n5,
    4:101–4:102

Indian Laburnum, 2:121n1

Indian liquorice, 3:565n1

Indian Oak, 2:464n3

Indian panic, *4:104*

Indian Reed, *3:180*, 3:181–3:185

Indicum
    Brasilianum, 4:146
    repens, 4:146

Indigo, 4:141, *4:142*, 4:143–4:147

Indigofera
    suffruticosa, 4:143
    sumatrana, 4:141n2
    tinctoria, 4:141n2, 5:79n4

Indjing, 4:113

Ingan Ingan, 3:393

Ingomaes, 4:373, 4:379

Injarro, 3:154

Ink plant, 5:83

Innapa, 3:250

Inocarpus
    edulis, 2:11n2, 2:85n2, 3:46n14,
        3:615n11
    fagifer, 1:455n1, 2:271n2,
        2:324n26, 2:391n7, 3:213n9,
        5:173n5, 5:205n22, 5:407n4

Inophyllum, 2:326
    flore octisido, 2:326

Inota inodien, 5:117

Intsia, 2:440*n5*
  bujuga, 1:348*n6,* 2:242,
    2:324*n24,* 2:389*n15,* 2:439*n2,*
    2:467*n8,* 2:468*n10,* 3:149*n3,*
    3:413*n2,* 3:475*n3,* 3:477*n13,*
    3:585*n1*
Intsji, 1:439
Inula helenium, 5:107*n1,* 5:545*n4*
Involucrum, 5:218, 5:219–5:220
  cusci, 3:297
Ipomoea
  aquatica, 4:461*n1*
  batatas, 4:369*n1,* 4:461*n2,*
    4:511*n3*
  coccinea, 4:466*n4*
  gracilis, 4:487*n1*
  indica, 4:483*n1*
  peltata, 4:477*n7*
  pes-caprae, 4:369*n4,* 4:413*n2,*
    4:478*n12,* 4:485*n1*
  quamoclit, 4:465*n1,* 5:571n4
  reptans, 5:43*n6,* 5:335*n3*
  rumphii, 4:477*n5*
Iridaceae, 4:60*n13*
Iridomyrmex, 5:231*n8*
Iripa, 1:449
Iris, 5:180*n10*
Iris
  florentina, 3:521, 3:521*n4*
  foetidissima, 4:33*n8*
  lutea, 4:33
Iron tree, 1:319, 1:320, 5:359
  large-leaved, 2:555
Ironwood, 2:87, 2:242, 2:324,
  2:631, 3:103, 3:149*n3,* 3:413,
  3:475, 3:477, 3:585, 5:201, 5:364,
  5:372, 5:563*n2*
  Ambonese, 2:*438,* 2:439–2:442
  bastard, 2:447, 2:*448,* 2:449
  common, 2:*438,* 2:439–2:442
  East Indian, 1:320*n21*
  Moluccan, 5:201*n8*
  mountain, 2:449
  taëwan, 2:447, 2:449
  Ternatan, 2:443, 2:*444,* 2:*445,*
    2:446

true, 2:429, 2:*430,* 2:431–2:432,
  2:*433,* 2:434–2:435, 2:*436,*
  2:437
Isatis indica, 4:141, 4:*142,* 4:143–
  4:147
Isca de Jaqua, 5:237
Ischaemum
  muticum, 5:25*n1*
  timorense, 5:25*n1*
Isidos plocamos, 5:364
Isis hippuris, 2:75, 5:387*n21,*
  5:389*n36,* 5:411*n1,* 5:430*n11*
Isis' hair, 5:364
Isora murri, 5:527
Isse Selu, 2:242
Isseputi, 3:16, 3:19
Issi Calappa, 1:193
Issipouan, 5:49, 5:*50,* 5:51, 5:*52,*
  5:53
Itam, 4:27
Itching plant, 4:419*n4*
Itelpou, 4:307
Itilat, 3:405
Itthion, 2:56
Itti Arealu, 3:25
Ivy, ground, 4:523*n2,* 4:525
Ixia, 5:30
Ixora
  chinensis, 3:360*n8*
  fulgens, 3:357*n1,* 4:522*n12,*
    5:532*n1*

J

Ja tsjau, 5:291
Jacca, 1:344
Jack fruit, 1:344*n14,* 1:350*n6*
Jack tree, 1:*342,* 1:343–1:345, 1:347
Jacqueira, 1:345
Jagan bajang, 4:159
Jager tree, 1:212*n21,* 1:215, 1:253,
  1:*254,* 1:255–1:261
Jagera serrata, 1:420*n26*

Jagerboom, 1:212*n21*, 1:256

Jagum, 4:115
  castela, 4:115

Jahaca, 1:273

Jajagua, 4:153

Jajama, 4:153

Jaka de Mato, 1:345

Jambis, 1:226

Jambo, 1:372
  alphuru, 1:385
  ayer puti, 1:377
  d'Agoarosada, 1:373
  d'Malacca, 1:373
  Portugal, 1:404
  sakelat, 1:385
  utan, 1:375

Jambolaan, 1:387

Jambolana, 1:387

Jambolano, 1:388

Jambolans tree, *1:386*, 1:387–1:388

Jamboli, 1:372

Jambolyns d' Juat, 1:387

Jambosa
  alphurica, 1:385
  aquea, 1:377
  calapparia, 1:371
  ceramica, 1:385
  domestica, 1:372
  litorea, 2:558, 2:559
  nigra, 1:375, 1:385
  silvestris
    alba, 1:379, 2:559
    parvifolia, 1:381

Jambosa nigra, 3:121

Jambu, 2:548
  ayer, 1:377, 1:385
    utan, 1:383
  clongcong, 1:371
  dayan, 1:372
  djene, 1:377
  itam, 1:375
  Laut, 2:559
  Pantey, 2:559
  sakkelät, 3:141
  soor, 1:469
  uer, 1:377

utan puti, 1:379
  white water, 1:377*n4*

Jambus, 1:333, 2:4, 2:21, 2:119,
  3:95, 3:141, 3:323, 5:468
  beach, 2:558, 2:559
  bearer, 1:469
  black, *1:374*, 1:375–1:376
  Ceremese, *1:384*, 1:385
  domestic, 1:369, *1:370*, 1:371–
    1:373
  forest, 1:375*n5*
  mountain, 1:381
  water, *1:374*, 1:377–1:378
  wild white, 1:379, *1:380*, 1:381–
    1:382

Jamo, 5:343

Jampadares, 1:348

Jampelong, 2:323

Jampiringan, 4:380

Jamtsja, 2:431

Jangang Ulang, 3:142

Japponse aardakers, 4:474

Jaqua papa, 1:345

Jaqueiro, 1:344

Jasmine, 4:139*n44*, 5:569
  Arabian, 2:303*n10*
  humile, 4:470

Jasminum, 3:258, 3:327, 3:359,
  3:511, 3:521, 3:559, 3:560, 3:561,
  3:573, 3:590
  Americanum, 4:466
  arabicum, 3:327, 3:558
  bifarium, 3:560*n20*
  litoreum, 3:560, 3:617, *3:618*,
    3:619
  Mexicanum, 4:192
  sambac, 2:129*n6*, 2:304*n11*,
    2:473*n2*, 3:30*n9*, 3:557*n1*,
    3:557*n4*, 4:62*n28*, 4:359*n4*,
    4:507*n5*, 5:433*n22*, 5:571n5,
    5:571n6

Jatropha, 3:287
  curcas, 3:339*n22*

Jatus, 2:452, 2:463–2:465

Jausi-Alinidi, 1:196

Java, 1:32–1:38

Java plum, 1:387*n1*

Javaanze
  boontjes, 4:401
  carwey, 4:219–4:220

Javanese cotton tree, 1:502, 1:503

Javanica, 3:221

Jealous plant, 5:111*n3*

Jets, 3:345

Jicama, 4:379*n1*

Jimsom, 5:549

Jimsonweed, 4:177*n2*

Job's tears, 5:41

Jujube tree, 2:*172*, 2:173, 2:174,
  3:250*n7*, 3:575
  beach, 2:*176*, 2:177–2:178

Jumble Beads, 3:565*n1*

Junceella juncea, 5:409*n1*

Juncia avellanada, 5:11

Juncus
  angulosus, 5:5
  indicus porosus, 3:641
  marinus, 5:410
  odoratus, 4:86
  quadratus, 5:5
  triangularis, 5:5

Juniper, 3:249

Juniperus chinensis, 2:88*n37*

Jussiaea
  erecta, 5:93*n1*, 5:94*n12*
  suffruticosa, 5:93*n1*

Justicia gendarussa, 2:333*n6*,
  3:293*n1*, 3:417*n3*, 3:569*n6*,
  3:619*n13*, 5:61*n13*, 5:95*n3*,
  5:471*n16*

K

Kacala, 4:45

Kadiri, 1:35

Kaë, 1:279

Kaempferia
  galanga, 3:320*n9*, 3:622*n9*,
    4:48*n6*, 4:73*n1*, 4:120*n7*,
    4:138*n28*, 4:331*n3*, 5:268*n5*,
    5:471*n18*
  pandurata, 4:71*n1*, 5:223*n11*

Kahali, 1:352

Kaida
  taddi, 3:442
  tsjerria, 3:451

Kaju-Mata-Buta Tree, 2:*358*,
  2:*359*–2:*361*, 2:*362*, 2:363

Kakae Mea, 2:595

Kakahir, 3:575, 3:577

Kakibele, 3:188

Kaku valli, 3:483

Kalan kebu, 5:136

Kalanchoe laciniata, 3:559*n17*,
  4:227*n1*, 4:258*n13*

Kalappa laut, 5:386

Kalbahaar, 5:357, 5:364, 5:369,
  5:*370*, 5:371–5:373
  batu, 5:419
  gajang, 5:407
  punja ackar, 5:365
  puti, 5:414
    lacki lacki, 5:409–5:410
    paraampuan, 5:414*n4*
  white, 5:411, 5:*412*, 5:413–5:417

Kalengi cansjava, 4:126

Kali, 5:315

Kaliatour wood. *See* Sandalwood

Kalk, 5:446

Kallape tree, 1:199

Kallar, 1:491

Kalong lamamam, 5:499

Kaly, 5:315, 5:317

Kam, 5:535

Kamamdre, 3:345

Kameels-hooy, 4:86

Kanawa, 2:344

Kangschin, 3:501

Kannekas de bugio, 3:675

Kannekens-kruyd, 3:673–3:676

Kapialun, 4:515

Kapihut, 3:477

Kappa kelengu, 4:372

Karandi, 4:506

Kareta tsjori valli, 4:517

Karkarren, 4:392

Kasjam tree, 3:524

Kateppeng, 4:237

Kathu karohiti, 5:313

Katjang, 3:615, 4:389, 4:403, 4:415,
5:289, 5:304
baly, 4:389
grass, 5:105
gunung, 5:280, 5:281–5:282,
5:282
idjo, 4:405n1
jappon, 4:474
kitsjil, 4:405–4:407, 4:409
laut, 4:415
mera, 4:401
puti, 4:401
sina, 4:383
tana, 4:474
tulis, 4:399
utan, 5:289

Katou-kadel, 3:426

Katsji, 4:528
kelengu, 4:338

Katskoppen, 5:446

Kattu
nurem kelengu, 4:358
tirtava, 4:256

Katu
katsjil, 4:506
kelangu, 4:351
peru, 4:392
tsjaka, 2:468
tsjetti pu, 4:204

Kaucki Tree, 2:289

Kauwas, 3:91

Kay kayl, 3:575

Kebacktsji, 5:535

Kedang, 4:6

Kehu, 5:245

Kekacktsji, 3:510

Kekatsji, 3:508

Kekel, 3:435, 3:442

Keker
ewan, 3:443, 3:445
leynulun, 3:442

moni, 3:446, 3:447
mountain, 3:453
wassi, 3:443

Kekin malona, 4:285

Kelady, 5:321
ajer, 4:287
alang, 4:285
annac, 4:287
ayer, 4:291, 4:292, 4:293
babi, 4:291
bulu, 4:287
candhati, 4:291
domestic, 4:285
fragrant, 4:287n17
gomuto, 4:287
sweet, 4:293
tauwar, 4:287
telloor, 4:287
wangi, 4:287
water, 4:285, 4:291, 4:292, 4:293
xula, 4:291, 4:293

Kelala Lacki, 2:353

Kelangissan, 3:215

Keli assapin, 4:339

Kelibessi, 2:528

Kellor, 1:483, 1:484, 1:485–1:486,
1:487, 1:488, 4:63, 4:326
gunong, 3:90, 3:135
laut, 2:497, 3:135, 3:277
utan, 2:497

Kellu, 5:317

Kemang tree, 1:334n9

Kembang sepatu, 3:219n16

Kenal, 3:91

Kenari tree, 1:202

Kenassal, 3:100

Kenay, 3:173

Kenehune, 3:24

Kering keringan, 4:232

Kernel sockam tree, 1:353

Kerswortel, 3:593

Kethoneth, 3:234

Ketmia, 2:334, 3:224

Khi, 1:398, 1:398n16

Kiamban kitsjil, 5:333, *5:334*

Kiar, 4:435

Kiate Tree, 2:*452*, 2:463–2:465

Kichiium, 4:435

Kikajon, 3:336, 3:338, 3:340

Kilahi, 2:481

Kilam, 2:49, 2:74

Kili kili, 4:237

Kimhong, 4:197

Kinamëan, 2:508

Kinar tree, 3:91, *3:92*, 3:93, 4:551

Kindjarong, 4:211

Kindra, 4:459

King's fruit, 5:525, *5:526*, 5:527

Kiokhoae, 4:200

Kion, 4:51

Kipper nut, 5:235*n29*

Kiras, 1:389, 1:391, 1:394

Kirganeli, 5:81

Kischa, 4:435

Kitania, 4:401

Kitnith, 4:401

Kitsjil, 3:108, 3:615
    parampuan, 3:257

Kittel, 3:234

Kleba, 3:451

Kleef-pruymen, 3:49

Kleene Geele Gom-boom, 2:591

Kleene praatjes, 3:631, 3:635–3:636

Kleinhovia hospita, 3:91*n1*, 3:115*n3*,
    4:326*n11*, 4:551*n6*, 4:582*n9*,
    5:140*n12*, 5:152*n21*, 5:189*n11*,
    5:201*n4*, 5:201*n7*, 5:242*n8*

Klitsji, 3:275, 3:623, 3:625, 3:626
    besaar, 3:629

Knotgrass, 4:584*n11*, *5:13*, 5:15,
    5:21, 5:21*n3*, 5:22

Koae tzayn, 5:129

Koal, 1:250

Koe Tsjandanam, 2:72

Koeskoes, 4:584, 5:23, *5:24*

Kofo sangi, 4:25*n6*

Kokerlingen, 4:556

Kokin, 4:27, *4:28*, 4:29–4:30
    mera, 4:27
    soffoheca, 4:572

Kondam pallu, 5:75

Kookernoot, 1:196

Kopsteen, 5:446

Kore Tree, 2:491, *2:492*, 2:493

Kota Ambon, 1:68–1:87

Koukoae, 4:447

Kraalsteen, 5:446

Krang gunong, 4:91

Krotje, 3:34, 3:41

Kuge, 4:477

Kukurang, 4:529

Kula, 4:6, 4:6*n37*
    abbal, 4:25
    bey, 4:23
    hatuan, 4:21
    keker, 4:11, 4:11*n30*

Kule, 1:429

Kullehur, 2:571

Kullur, 1:351, 1:356

Kullus utan, 1:357

Kuntsu, 4:73

Kutthonet, 3:234

Kyllinga monocephala, 5:15*n1*

Kyó, 4:169

L

La Renette, 4:155

La-hua-sheng, 4:473*n5*

Laar tree, 3:30

Labi labi, 5:345

Lablab purpureus, 3:154*n7*, 4:391*n1*,
    4:393*n1*, 4:395*n6*

Labo
    ambon, 4:429
    castela, 4:429
    frangi, 4:427
    panjang, 4:427
    puti, 4:427, *4:428*, 4:429

Labrusca molucca, 4:518, 4:519

Labuttan, 3:108

Labutun, 1:463

Lacca, 3:498, 3:499–3:502
  alba, 4:196
  ambon, 4:196
  api, 4:195
  coccinea, 4:195
  cuning, 3:507
  frutex, 4:196
  garo, 3:507
  herba, 4:194, 4:195–4:197
  lignum, 3:498, 3:499–3:502,
    4:196
  purpurea, 4:196

Lachryma Jobi Indica, 4:98, 4:100n5,
  4:101–4:102

Lacki lacki, 5:99

Lackiahat, 3:463

Lacong, 4:515

Lactaria
  herba, 2:370
  salubris, 2:385, 2:386

Lactuca, 4:228n4, 5:73n5, 5:107
  aquarum, 5:346
  marina, 5:346
  sativa, 4:551n2

Lada, 4:186
  andjing, 3:551
  pandejang, 4:315
  tschili, 4:186

Ladja
  lobbe, 5:267
  lobu, 5:274

Ladle shrub, 3:329, *3:330,* 3:331–
  3:333

Lady fern, 4:270n30

Lady's slippeer, 5:179n3

Lagan, 3:21

Lagansa, 4:230, 4:233–4:234
  alba, 4:233
  rubra, 4:233

Lage lage, 5:133, 5:136

Lagenaria
  leucantha, 4:285n3
  siceraria, 4:425n1, 4:426n11,
    4:427n1

Lagerstroemia
  indica, 5:579n1
  specisa, 4:161n4

Lagondi shrub, 3:263, 3:337
  beach, *3:259, 3:260,* 3:261
  common, *3:255, 3:256,* 3:257–
    3:258
  laut lacki lacki, *3:259, 3:260,*
    3:261

Lagondium
  litoreum arborescens, 3:259, 3:260,
    3:261
  vulgare, 3:255, 3:256, 3:257–
    3:258

Lagulu, *5:49, 5:50, 5:51, 5:52,* 5:53

Lagundi, 1:279

Lahac, 3:605

Lahang, 1:273

Lahar Tree, *2:482,* 2:483

Laharong, stone, 2:483

Laharus, 2:482, 2:483

Laicou, 5:136

Lalamut, 5:353

Lalan, *5:32, 5:33* 5:34
  bitauer, 2:591

Lalary Wood, *2:599, 2:600,* 2:601–
  2:604

Laleri, 4:485

Laletsjong, 1:463

Lalia, 1:466

Laluccu, 1:196

Lambuto, 2:370, 2:371

Lami, 4:312
  huta, 4:312
  java, 4:307

Laminaria, 5:337n6

Lammut, 1:446
  abbal, 1:449

Lamp-rosen, 3:461

Lamp-Roses, 3:461n7

Lampaos, 4:497

Lampatan, 4:497

Lampery, *3:418*, 3:419

Lampuin, 4:39

Lampujang, 3:294, 5:223, 5:310

Lampujang, 4:48, 4:54, 4:57*n2*,
  4:67, 4:68, 4:84
  besaar, 4:39
  kitsjil, 4:39
  matta mingil, 4:39
  wangi, 4:39

Lampujum
  major, 4:37, *4:38*, 4:39–4:40
  minus, *4:38*, 4:39–4:40
  silvestre amarum, 4:42

Lamtassy, 2:627

Lances christi, 5:292, 5:295

Lanck-was, 4:35

Land Casuaris Tree, 2:570

Land Ohupaë Tree, 2:280

Langa langa, 5:65

Lange peper, 4:315

Langir, 3:413, *3:414*, 3:415

Langondi, 1:279*n34*, 4:389

Langondium vulgare, 4:182*n41*

Langsa tree, 1:204*n27*

Languas galanga, 4:30*n17*

Lanha, 1:196

Lani tree, *3:124*, 3:125–3:126

Lanius, *3:124*, 3:125–3:126

Lanpujum
  silvestre minus, 4:41–4:42
  utan kitsjil, 4:41

Lanquas, 4:31, 4:33, 4:35, 4:54,
  4:77, 5:317
  bocassan, 4:36
  globba, 5:267, 5:274
  kitsjil, 4:35, 5:267
  lacki lacki, 5:271
  Malacca, 4:45
  mera, 4:79
  tsjendana, 4:35

Lansa
  domestic, *1:422*, 1:423–1:424
  gunong, 1:428
  mountain, 1:428, 5:493
  utan, 1:425
  wild, 1:425, *1:426*

Lansac, 1:424

Lanse, 3:51

Lanse Tree, 2:89, 2:427, 2:427*n2*

Lansia, 5:561

Lansium, 1:424
  domesticum, 1:204*n27*, 1:228*n41*,
    1:242*n12*, 1:389*n5*, 1:423*n1*,
    1:506*n11*, 2:89*n3*, 2:102*n11*,
    2:119*n3*, 2:177*n3*, 2:207*n4*,
    2:427*n2*, 2:434*n21*, 2:460*n2*,
    2:485*n2*, 2:555*n3*, 3:16*n10*,
    3:51*n3*, 3:71*n8*, 3:389*n2*,
    3:491*n2*, 3:531*n2*, 3:613*n3*,
    5:301*n6*, 5:473*n3*, 5:476*n4*,
    5:499*n2*
  montanum, 1:428
  silvestre, 1:425
  silvestri, 2:493

Lantanier tree, 1:266

Lantebu, 3:209

Lanze tree, 3:71

Lapathum hortense, 4:229

Lape Lape, 2:552

Lapia, 1:299
  ihur, 1:298
  luli-uwe, 1:296
  macanaru, 1:298
  molat, 1:298
  shur, 1:296
  tuni, 1:296, 1:298

Laportea amplissima, 2:387*n3*,
  3:175*n1*, 3:175*n3*, 3:423*n1*

Lappago
  amboinica, 5:113
  laciniata, *5:110*, 5:113–5:114
  latifolia, 5:113
  nussalaviensis, *5:110*, 5:113–5:114

Lapussi, 3:239

Laquanqua hujtl, 1:503

Laranza Tree, 2:168

Large marbles shrub, 3:627, *3:628*, 3:629–3:630

Larike, 1:59–1:62

Larin, 4:27

Lassa, 1:424
  suru, 1:425

Lassata, 1:424

Lassi Batu, 2:479

Lassiaal, 3:463

Latak, 5:346

Late
  besaar, *5:32, 5:36,* 5:37–5:39
  kitsjil, *5:32, 5:36,* 5:37–5:39

Lathyrus, 2:359, 3:283
  tuberosus, 3:563*n2,* 4:377*n2,*
    4:474*n10*

Latifolia, 3:4

Lau
  lassi, 1:473
  takka, 1:420

Lauassin, 2:589

Laueweul, 2:570

Laulau tomma wanga, *5:26,* 5:29–
  5:30

Laun maun, 2:555, 3:615

Launia sanaa, 3:427

Laupa, 5:5
  layn, 5:5

Lauraster
  amboinicus, 2:89, *2:100,* 2:101–
    2:102, *2:535*
  minor, 2:102

Laurel, 2:3, 2:4, 2:93, *2:94,* 2:95–
  2:98, 3:173, 3:300, 3:359, 3:521,
  5:583–5:584
  Leytun, *2:100,* 2:101–2:102
  wild Ambonese, *2:100,* 2:101–
    2:102

Laurifolia, 1:391

Laurus
  japonica, 5:583–5:584
  nobilis, 5:505*n3*

Lauwassin, 2:511

Lavandon, 4:33

Lavandula stoechas, 5:555*n1*

Lavn-mahu, 1:385

Lawassa Malacca, 4:77

Lawassi malacca, 5:268

Lawat seru, 5:152*n21*

Lawos, 4:35

Lawson, Isaac, 1:142

Lawsonia inermis, 1:323*n5,* 1:455*n2,*
  2:527*n5,* 2:639*n4,* 3:162*n7,*
  3:253*n30,* 3:499*n2,* 4:138*n29,*
  4:195*n1,* 4:196*n8,* 4:507*n4,*
  4:535*n16,* 5:155*n8*

Leaf of Queens, *3:370,* 3:371–3:372

Leaf of Tongues, *3:466,* 3:467–
  3:469

Leca leca, 4:39

Leco bulu bulu, *5:40,* 5:43–5:45

Lectau, 4:405

Leea
  aculeata, 1:275*n3,* 2:451*n7,*
    3:351*n1,* 3:421*n4,* 3:423*n3*
  aequata, 2:511*n6,* 3:355*n1,*
    4:472*n18*
  indica, 3:355*n5*

Leeker, 4:303

Leeks, 4:410*n9*

Leimis Tree, 2:155, *2:156, 2:157*

Leiopathes glaberrima, 5:377*n5*

Leja, 4:51

Leko, 4:320

Leleba, *3:180,* 3:181–3:185, 3:189

Lelou, 5:548

Lemmack Daging, 2:261–2:262

Lemna minor, 5:333*n2*

Lemo
  lemo, 2:517
  tenuis, 2:155

Lemon, 2:61, 2:143, *2:144, 2:145,*
  2:367
  balm, 4:257*n5*
  banda, 2:142
  belly, *2:146,* 2:147–2:149
  bonting, 2:142

bras, 2:143
buleyn, 2:159
calucku, 2:140
cambar, 2:152
Capas, 2:155
Carbou, *2:150*, 2:151, 2:152,
  2:157
carbou, 5:304
Cassomba, 2:140
Colte, 2:171
Curamas, 2:149
Golden, 2:149
gula, 1:509
Itam, 2:141, 2:163, *2:164*, 2:165,
  2:179
Jabba, 2:151
Java, 2:145, 2:155
Maas, 2:162
manis, 2:263
  tsjina, 2:168
Maritsja, 2:165
Martin, 2:147, 2:151, 2:167
Martinjo, 2:141
nipis, 1:398, 2:155, *2:156*, 2:159,
  2:162, 2:165, 2:173, 2:179,
  2:183, 2:467, 2:513, 2:570,
  3:105, 3:107, 3:573, 3:607,
  3:619, 4:291, 4:347, 5:326,
  5:415, 5:444, 5:493, 5:531,
  5:531*n9*, 5:569
Padangh, 2:152
Papeda, 2:157
  wild, *2:150*, 2:151–2:152
Papua, 2:149
pumpkin, 1:341*n38*
Purrut, 2:148
Songalea, 2:168
Spanish, 2:147, 2:148, 2:153
Suassi, 2:171
Sussu, 2:143
swangi, 2:153, *2:154*, 3:118,
  3:134, 3:192, 3:275, 3:535,
  5:53, 5:101
Tinagara, 2:159
Tsjina, 2:169
Utan, *2:161*, 2:162
water, 2:406*n7*
wood, 3:501*n24*
Lemun Manus Besaar Tree, 2:169

Lenat, 1:473
  frangi, 1:473
Lenni, 1:196
Lens
  palustris, 5:333
  phaseoloides, 3:478
Lenticula marina, 5:349
Lentinus
  djamor, 5:240n7
  regium, 5:233*n4*
  sajor caju, 5:240n1
  subnudus, 5:240n2
  tuber regium, 5:233*n2*
Lepidagathis rumphii, 3:295*n1*,
  5:99*n2*
Lepidium latifolium, 4:183*n5*
Lepiniopsis ternatensis, 2:577*n1*
Lepisanthes, 2:593*n1*
Leptosperamum flavescens, 1:277*n17*,
  2:111*n1*, 5:133*n3*
Lerio manobi, 4:474
Lethrinus
  amboinensis, 3:593*n2*
  olivaceus, 3:593*n2*
Lettuce, 1:495*n2*, 4:228*n4*, 5:107*n5*
  water, 4:243*n4*, 5:333–5:334
Leucas
  lavandulifolia, 5:73*n1*
  zeylanica, 5:73*n1*
Leucojum, 3:269, 3:269*n3*
  luteum, 2:313
Leut, 3:459
Levisticum
  indicum, 4:217
  officinale, 4:217*n1*
Lewani, 5:489
Lewer, 1:473
Leyitir Puti, 2:300
Leytewan, 3:443
Leytimor Wersingin, 3:314
Leytun, *2:100*, 2:101–2:102
Liada, 3:103
Lian, *3:470*, 3:471

Liat sahari, 5:309

Libe halun, 4:519

Licorice shrub, 3:84

Licuala, 1:249, 1:250, 1:266, 5:544
  rumphii, 1:249*n8*, 2:88, 2:88*n32*,
    3:664*n10*, 4:262*n10*
  spinosa, 1:249*n8*

Lida
  andjing, 5:65
  boyaya, 3:329, 4:221

Liemis Tree
  golden, 2:158, 2:159, 2:160,
    2:161, 2:162
  Maduran, 2:159, 2:160, 2:162

Lien, 5:321

Light of the World, 3:266

Ligneum corneum, 2:505

Lignum
  aloës, 2:49, 3:501
  aquatile, 3:422, 3:423
  citrinum, 1:277, 3:417
  clavorum, 2:589, 2:590, 2:591
  colubrinum, 2:179, 2:180, 2:181–
    2:182
  corneum, 2:503, 2:504, 2:505,
    2:507
    angustifolium, 2:503
  emanum, 2:488, 2:489–2:490,
    5:503, 5:503*n6*
  equinum, 2:540, 2:541–2:542
  eurinum, 2:519
  ferreum
    verum, 2:431
    vulgare, 2:440
  guajacum, 2:414, 2:431, 4:499
  leve, 2:534, 2:535–2:536, 2:537
    alterum, 2:538, 2:539
  moluccense, 3:347
  momentaneum, 3:64, 3:65
  moschatum, 2:62
  mucosum, 3:143, 3:144
  murenae, 3:135
  murinum, 2:496
  Papuanum, 2:83–2:84
  salis, 2:507
    minus, 2:515
    primum, 2:508

sappanum, 3:268, 3:269–3:272

scholare, 2:372, 2:373–2:375,
  3:524

suffitus, 2:79

Timorense, 2:179, 2:180, 2:181–
  2:182

vinosum, 5:507

viperinum, 4:560

vitae, 2:496*n11*

Ligularia, 3:329, 3:330, 3:331–
  3:333
  lactea, 2:62, 2:189, 2:361, 2:379,
    3:286, 3:330

Ligusticum acutilobum, 4:217*n2*,
  4:217*n5*

Ligustrum, 1:323, 3:249, 3:250,
  3:252, 3:253, 3:308
  aegyptium, 3:250
  indicum, 3:255, 3:256, 3:257–
    3:258
  vulgare, 3:273*n2*

Lilac
  Malabar, 3:327
  wild, white, 3:552, 3:553–3:554

Lilium
  convallium, 2:23, 2:195, 2:581,
    2:593, 3:557
  indicum, 5:305, 5:306, 5:309–
    5:310

Lily
  Indian, 5:306, 5:309–5:310
  Indies, 5:205*n23*
  water, 5:163*n8*, 5:319, 5:320,
    5:321–5:324, 5:325–5:329

Lily of the Valley, 3:557*n3*

Lima turi, 5:503

Limax saxea, 5:444

Lime Leaf, 3:407, 3:408, 3:409

Limestone, marine, 5:445–5:448

Limnanthemum indicum, 5:326*n4*

Limnophila
  aromatica, 5:171*n27*, 5:291*n1*
  rugosa, 3:372*n13*, 5:293*n1*

Limo
  acidissima, 2:62*n1*
  agrestis, 2:151–2:152

crispus, 2:149
ferus, 2:153
martinicus, 2:147
praegnans, 2:142
pultarius, 2:151
taurinus, 2:152
tubreosus, 2:146, 2:147–2:149
unguentarius, 2:149
ventricosus, 2:148
Limon
  Crispus, 2:165
  Maas, *2:158*, 2:159
  Martyn, 2:155
  Pouleron, 2:171
  Swangi, 2:153, *2:154*
  Titigela, 2:153, *2:154*
Limondoce Tree, 2:168
Limonellus, 2:155
  angulosus, 2:162
  aurarius, 2:158, 2:159, *2:160*,
    2:161, 2:162, 2:609
  aureus, 3:321
  litoreau, 3:508, 3:510
  littoreus, 2:177
  Madurensis, 2:160, 2:162
Linaoo wood, 2:71
Linden, 3:91, 3:131
Lindernia crustacea, 4:533*n2*,
  5:129*n6*
Linen shrub, *3:376*, 3:377–3:379
Lingo, 2:366
  Parampuan, 2:315
  Puti, 2:315
Lingoa, *2:312*, 2:313–2:318, 3:103,
  3:127, 3:378, 3:560
  Bonga, 2:315
  Meera, 2:315
  stone, 2:319, *2:320*
Lingoo, 1:457, 2:384, 2:422, 3:23,
  3:499, 3:500, 3:532, 3:539, 4:326,
  5:476, 5:561
  Batu, 2:63, 2:319
  red, 2:313–2:314, 2:315
  stony, 2:313
  Tsjampagga, 2:315
  white, 2:314, 2:315
  wood, 2:74, 2:79, 2:85

Lingoum
  molle, 2:315
  rubrum, 2:315
  saxatile, 2:319, *2:320*
Lingua
  canina, 3:333
  de vacca, 4:262
Liparis condylobulbon, 5:205*n21*
Liplap, 1:193
Liquidamber, 2:236*n14*
  orientalis, 2:83*n1*, 2:87*n28*,
    2:597*n3*
Lire, *5:56*, 5:57–5:58
  kitsjil, 5:58
  lire, 5:60
  papoua, 5:58
  petola, 5:58
Litchi chinensis, 5:235*n17*
Lithodendrum, 5:357
  calcarium, *5:428*, 5:429–5:433
    erectum, 5:429
    sessile, 5:429–5:430
  digitatum, 5:430
  litoreum, 5:430
  saccharaceum, 5:435
Lithophaga teres, 5:445*n4*
Lithospermum, *5:40*, 5:41
Litsea, 3:67*n4*
  rumphii, 2:539*n1*
  stickmanii, 2:535*n1*
Little Devil Pots, 3:674
Little red bell, 5:543
Little time flower, 4:192*n11*
Liung tsi tschau, 4:221
Livistonia rotundifolia, 1:247*n1*,
  1:300*n38*, 5:544*n1*
Lo-ca-tsay, 5:171, 5:171*n18*
Loacca, 1:194
Loanda, 2:603
Lobus
  brasilianus ingens, 3:476
  cartilagineus, 3:481, *3:482*, 3:483
  echinodes, 3:629, 3:629*n10*
  litoralis, 3:481, *3:482*, 3:483
  machaeroides, *4:386*, 4:387

membranaceos, 3:633
quadrangularis, 4:381, *4:382*

Lodoicea
maldivica, 1:204*n32*, 2:75,
2:404*n39*, 5:126*n17*, 5:383*n1*
sechellarum, 5:357*n10*, 5:383*n1*

Loebessen, 2:437

Logwood, 3:270*n18*, 3:272*n36*,
3:501*n24*

Lohussul, 2:499, 2:505

Loinhuan, 3:211

Lojang, 5:364

Lolan, 3:270, 3:276
beach, 3:273
bean, 3:273
Key, 2:293
laynuhi, 3:273
pantey, 3:273
Waran, 2:589

Lolaro, 2:601

Lolaru
Tomma Banga, 2:508

Lolay Tree, 2:555, *2:556*, 2:557

Lolin Tree, 2:427

Lolium
perenne, 5:27*n2*
temulentum, 5:35*n3*

Loloro, 4:485, *4:486*

Lolowararan Tree, 2:367

Lomba, 5:258, 5:259

Lommu, 1:460

Lompa, 4:528

Lonchitis
amara, 5:136
amboinica, 5:132, *5:133*, *5:134*,
5:135–5:137
mucosa, 5:136–5:137
pilosa, 5:136
recta
major, 5:133
minor, 5:133, 5:135
saguaria, 5:136
volubilis, 5:135, 5:136, 5:137

Long-lived Wood, 3:263

Lontar, 1:253–1:261
lacki lacki, 1:261
utan, 1:265, 1:269
wild, 1:263, *1:264*, 1:265–1:267,
1:268, 1:269

Lontarus
domestica, 1:256
femina, 1:261
mas, 1:261
silvestris, 1:265, 1:272
altera, 1:269

Loosestrife, 3:293*n2*

Lophopyxis
maingayi, 3:593*n1*
pierrei, 3:593*n1*

Loranthus rumphii, 2:222*n25*,
3:569*n4*, 3:615*n13*

Loro Waran, 2:589

Lositti, 3:319
salu uhi, 3:321

Lossa lossa, 4:534

Lottu
lottu, 5:339
metten, 5:340
puti, 5:340

Lotus
aegyptian, 5:327
blue, 5:326*n14*
white, 5:326*n12*

Lotus, 4:288*n24*
aegyptia, 4:288, 5:322, 5:326,
5:327
urbana, 4:271

Lou
furu, 3:188
lite, 2:385
lou Idje, 3:209

Louw Louw, 3:188

Love herb, 5:67, 5:80

Love plant, 4:268*n8*

Lowan Cambodja, 2:379

Lucuala rumphii, 2:383*n9*

Lucuma lucida, 3:105*n1*

Luffa
  acutangula, 4:443*n1*
  aegyptiaca, 4:437*n1*, 4:437*n6*,
    4:445*n1*, 5:367*n22*
  cylindrica, 4:25*n15*
Luffe luffe, 4:209
Luha, 3:407
Luhu, 1:273
Luinut, 5:167
Luisendoder, 3:388
Luisia confusa, 5:199*n13*
Lulana toa, 3:365
Luluhan, 4:209
Lumis, 1:325
Lumut laut, 5:337
Lunga tambang, 3:319
Lunt tree, 2:598, 3:115*n1*
  black, 3:113
  white, 3:109, *3:110,* 3:111
Lupinus, 3:83, 3:265, 3:265*n7*
Lussa radja, 5:483, *5:518,* 5:519–
  5:520
Lussu layn, 3:273
Lussukey Tree, 2:293
Lutto Pya, 2:559
Luttun, 1:372
Lutu, 4:336
  castela, 4:370
Lux mundi, 3:264, 3:265–3:267
Lycium
  indicum, 1:234, 3:575
Lycoperdon, 5:251*n3*
Lycopersicon esculentum, 4:169*n8*,
  4:455*n1*, 4:457, 5:529*n4*
Lycoperson bovista, 3:134*n14*
Lycopodium
  cernuum, 5:165*n1*
  nummularifolium, 5:173*n3*
  phlegmaria, 5:173*n1*
Lydoriam e bengala, 1:506
Lygodium
  circinnatum, 1:214*n31*, 5:141*n2*
  scandens, 5:141*n10*

Lying-leaf, 3:314
Lyitohu, 2:374
Lysimachia, 3:293
Lysimachia, 5:93, 5:93*n2*, 5:94

M

Maa, 1:273
  kakiri, 5:122
Maagden, 4:419
Maagden-lot, 3:300
Maagdenkruid, 4:420
Maäke, 3:549
Maandya, 1:345
Maanuno, 1:265
Maba buxifolia, 2:289*n8*, 2:409*n1*,
  5:489*n2*, 5:491*n4*
Macanalun, 1:298
Macaranga
  hispida, 3:133*n3*
  involucrata, 3:133*n1*, 3:133*n2*,
    3:321*n7*, 3:375*n8*
  mappa, 3:81*n1*
  tanarius, 2:546*n9*, 3:117*n5*,
    3:118*n19*
Macaraquean, 3:435
Macassaarze Gift-boom, 2:399–
  2:400
Maccabuhay, 4:245
Maccafuta, 2:363
Maccalium, 1:362
Maccanunut, 1:265
Maccassarese metrosideros, 5:468
Maccondou, 3:55
Machilus, 2:529, 2:530, 2:531,
  2:532, *2:533*
Mackilan, 5:575
Mackou, 5:83
Maclura
  cochinchinensis, 3:505*n2*
  javanensis, 3:271*n22*
Macropiper, 4:314, 4:315–4:317
Macuer plant, female, *5:254,* 5:255
Macuer plant, male, *5:254,* 5:257

Macueru lacki lacki, 5:254, 5:257

Macuerus
  faemina, 5:254, 5:255
  mas, 5:254, 5:257

Maculosum, 2:59

Mada kacka, 3:468

Madi, 1:226

Madicapo, 5:108

Madja, 1:506
  bim-wila, 1:506
  can, 2:14
  carbou, 1:509
  gahu, 2:363
  pahit, 1:36
  pissang, 1:507
  pu, 3:560
  telloor, 1:506

Madjong utu, 1:313

Madman's herb, 4:180

Madorius, 5:512, 5:513–5:515

Madreporaria, 5:449n1

Maesa tetrandra, 3:389n1

Magic plant, 5:57

Magic rope, 4:510

Magician's stone, 5:423

Magnolia, 1:401, 2:303n10,
  2:309n1

Magori tunkin, 5:571

Magudaris fruticescens, 3:309

Mague Verdadeiro, 2:603

Mahecul, 2:473

Mahoa pantey, 3:481

Mahuëwan, 3:78

Mahumäa, 3:151

Mail-Anschi, 3:254

Majana
  alba, 4:251, 4:252, 4:253
  aurea, 4:259
  bussu, 5:77
  foetida, 5:74, 5:77
  rubra, 4:251, 4:252, 4:253
  utan, 5:77

Majani, 4:251, 4:252, 4:253

Majapahit, 1:37

Majassa, 1:265

Majurat, 2:511

Makakiri, 5:140

Makassar, 1:44

Makellun halaluno, 3:297

Makilan, 1:39, 2:529, 2:530, 2:531,
  2:532, 2:533, 3:67

Makka, 1:273

Makui, 5:267

Makuy Malacca, 4:77

Mala
  elengi, 1:478
  mala, 4:229

Mala aurantia, 2:168
  Sinensia, 2:168

Malabar convolvulus, 4:463

Malabar conyza, 5:64

Malabar Plum Tree, 2:290, 3:49

Malabar Prune Tree, 1:376

Malabar Tree, 2:290

Malabathrum, 3:421, 4:320, 4:323

Malacca, 1:36
  pela, 1:405
  schambu, 1:373, 1:381

Malahjuur, 3:34

Malahoul, 5:105

Malahuol, 3:37

Malahuur, 2:437

Malaleuca cajuputi, 3:61n5

Malaparius, 3:102, 3:103–3:104

Malastoma, 3:429

Maling maling, 3:351

Malisa camul, 4:315

Mallotus tiliifolius, 3:131n1, 3:407n6

Mallow, 3:217n1

Malowassi, 5:489

Malum
  aruanum, 5:565, 5:566
  Assyrian, 2:140, 4:6, 4:6n24
  aureum, 2:168
  citrium, 2:143, 2:144, 2:145,
    2:147
  granatum littoreum, 1:277

hespericum, 2:147, 2:168
indicum, 2:174
insanum, 4:171
medicum, 2:147
orthomasticum, 2:155, 2:156,
   2:157
punicum, 2:136, 2:137–2:138

Malva, 3:133, 3:243
   arborescens, 3:223
   fruticescens, 3:221
   hortensis, 3:223
   indica, 3:247
   rotundifolia, 1:323n5
   sylvestria, 3:217n1
   visco bastardo, 3:227

Malvaceae, 2:173n4, 3:217n1

Malvae, 2:331

Malvas, 5:85

Mamanira shrub, *3:392,* 3:393
   white, *3:394,* 3:395

Mambu, 3:181, 3:193, 3:216

Mamelen, 2:467n5, 2:561, 3:52

Mamina angustifolia, 2:377

Mamori, 5:559

Mampelaan, 1:325, 1:327

Mamura, 1:417

Mamussut, 3:399

Mancudu, 3:51

Manda kacki, 5:517

Mandackkacki, 3:327

Mandarsah's Tree, 2:17

Mandaru Valli, 3:469

Mandihoca, 4:358

Mandragora, 5:545

Mandragora officinalis, 5:115n9,
   5:545n1

Mandragora officinarum, 4:179n17

Manfigera foetida, 3:413n4

Manga, 1:323, *1:324,* 1:325, *1:326,*
   1:327–1:329, 1:379, 2:381, *3:14,*
   3:15–3:16, *3:17,* 3:18, 5:180,
   5:181n28
   bacon, 1:325
   bangay, 1:325

barera, 1:325
berabu, 1:327n25
brava, 1:327, 2:279, 2:280, *2:368,*
   2:369–2:371, 2:385, 2:391,
   3:325
buzjo, 1:325
calappa, 1:325
dading, 1:325
daki, 1:325
dodol, 1:325
domestica, 1:327
flesh, 1:325
foetida, 1:333, 1:461, 3:105,
   3:107, 3:413, 5:262
kayr, 1:325
minor, 1:325
monkey, 1:325
simiarum, 1:325
stinker, *1:332,* 1:333–1:334
sylvestris, 1:331
utan, 1:331
wild, *1:330,* 1:331, *2:368,* 2:369–
   2:371
woït, 3:615

Mangelicaan, 4:205

Mangi Mangi, 2:503, 2:582, 3:7,
   3:119, 3:125, 3:169, 3:207, 3:436,
   3:543, 3:631
   Bonga, *2:640,* 2:641
   candle-bearing, *2:610,* 2:611–
      2:612, *2:613,* 2:614
   cheese-shaped, 2:615, *2:616,*
      2:617–2:618, *2:619,* 2:620,
      2:621
   clove-shaped, *2:628,* 2:629
   fingered, 2:607, *2:608,* 2:609
   flower-bearing, *2:640,* 2:641
   forest, 2:507
   Gunong, *2:636,* 2:637
   horned, 2:625, *2:626,* 2:627
   Iron, 2:631, *2:632,* 2:633
   mountain, 2:507, *2:636,* 2:637
   Padomara, 2:617
   Perampouan, 2:601
   purslane-leaved, *2:642,* 2:643
   small, 2:605, *2:606*
   tall, 2:599, *2:600,* 2:601–2:604
   true, 2:507n1
   Tsjencki, *2:628,* 2:629

Utan, 2:506, 2:507–2:508, 2:636, 2:637
uti, 2:623
white, 2:622, 2:623–2:624
wild, 2:506, 2:507–2:508
Mangi mangi, 1:389, 1:454, 4:549, 4:553
Mangifera, 1:327
  caesia, 1:334*n9,* 5:262*n11*
  foetida, 1:333*n1,* 1:334*n9,* 1:353*n17,* 2:79*n3,* 2:391*n6,* 3:105*n3,* 5:111*n2,* 5:501*n2*
  indica, 1:323, 1:323*n1,* 1:324, 1:325, 1:326, 1:327–1:329, 1:331*n1,* 2:268, 3:325*n2,* 4:11*n22,* 4:551*n7,* 5:150*n8,* 5:181*n28,* 5:384*n14,* 5:473*n4,* 5:501*n3,* 5:573*n2*
  laurina, 1:325*n23*
  minor, 1:323*n1,* 1:325*n20*
  rumphii, 1:329, 5:501*n1,* 5:501*n4*
  taipan, 1:331*n3*
Mangiium caryophylloides, 2:605
Mangiriquam, 4:205
Mungium
  album, 2:582, 2:599
  arcuatum, 2:612, 2:614
  candelarium, 2:599, 2:603, 2:604, 2:614, 3:455
  caryophylloides, 2:625, 2:628, 2:629
  caseolare, 2:519, 2:547, 2:599, 2:604, 2:612, 2:615, 2:616, 2:617–2:618, 2:619, 2:620, 2:621, 3:118, 4:551, 5:193, 5:195, 5:211
    rubrum, 2:618, 2:621
  celsum, 2:599–2:604, 2:607, 2:614
  corniculatum, 2:625, 2:627, 2:639
  digitatum, 2:599, 2:607, 2:608, 2:609
  ferreum, 2:411, 2:599, 2:631, 2:632, 2:633
  floridum, 2:640, 2:641
  fruticans, 2:582, 2:599, 2:604, 2:612, 2:624, 2:625, 2:639
    corniculatum, 2:627
    primum, 2:627
  legitimum, 2:599

  minus, 2:599, 2:605, 2:606, 3:271
  porcellanicum, 2:643
  silvestre, 2:506, 2:507–2:509, 2:511, 3:350
  terrestre, 2:505
  waccat, 2:519
Mangle Tree, 2:603, 2:604
Mangostan, 1:389, *1:390,* 1:391–1:392, *1:393,* 1:394, 5:463
  utan, 1:394
Mangostana, 1:389
  celebica silvestris, 1:394
Mangrove, 1:454*n12,* 2:38*n7,* 2:601*n5,* 2:601*n9,* 2:602*n17,* 2:603*n23,* 3:119*n9,* 3:169*n3,* 3:207, 3:433*n1,* 3:433*n4,* 3:436*n28,* 3:631*n7,* 4:545*n3,* 4:553*n4,* 5:140*n11,* 5:149*n2,* 5:150*n9,* 5:181*n43,* 5:196*n22,* 5:229*n4,* 5:311*n3,* 5:313*n9*
Manguba, 4:420
Mangue, 3:7
Manihot esculenta, 4:303*n9,* 4:304*n13*
Manija coua, 4:60
Manila, 3:219
  copal, 2:263*n1*
Manilhas Flower, 3:327, *3:328*
Manilkara, 2:437
Manipa, 1:38
Manoa papuwa, *1:399n4*
Manoapa puwa, 1:399
Manora flower, *3:556,* 3:557–*3:561*
Manto de nuestra senhora, 4:288
Manu Caloway, 3:71
Many-Colored Tree, *2:634,* 2:635
Manytsjadi, 3:84
Maöha, 4:395, 4:417
  utan, 4:419
Mapania macrocephala, 5:38*n13*
Maple, 2:265
  field, 4:447*n4*
  hedge, 4:447*n4*
Maramia, 3:24

Maranga mappa, 2:463*n*2

Maranta, 3:214, 3:216
  galanga, 4:30*n17*

Maravillas del Peru, 4:192

Marboo, 2:440

Mardeni, 4:196

Mare's tail, 5:414*n4*

Margosa, 4:447, 4:531

Margozero, 4:463

Maria Tree, 2:323*n14*

Marigold, 2:343

Maringa oleifera, 3:254*n38*

Marisa ahuë, 4:316

Maritsja laut, 3:542

Marmeleira, 1:506

Marocca, 3:73, *3:80*, 3:81
  nonau, 3:78, 3:81
  ternatan, 3:77, 3:78

Maroy, 5:267

Marrubium, 5:73*n3*
  album, 4:251, 5:80
    amboinica, 4:254, 4:257–4:258
  rubrum, 4:255
  vulgare, 4:257*n2*

Marshmallow, 4:127*n32*

Martahul, 1:277, 2:582, 2:591

Marurang, 3:365

Marvilje plant, *4:190*, 4:191–4:193

Massaan de Mato, 2:186

Massoy, 2:89–2:92, 4:63, 5:469,
  5:483

Massussu, 2:168

Masterwort, 3:301*n4*

Mastocarpus klenzeanus, 5:345*n8*

Mataram, 1:37

Mates de india pardo, 3:629

Matojo, 3:629

Matricaria, *4:198*, 4:199–4:201,
  4:199*n2*, 4:472
  recutita, 4:469*n3*

Matsjatam, 3:528

Matta
  auwe, 2:300
  ayn, 2:300
  cuttu, 3:388
  huli loumuri, 3:411
  huri, 2:363, 2:639
  kakiri, 5:136–5:137
  mangjil, 4:67
  matta, 4:569
  pannas, 5:51, 5:51*n6*, 5:53
  udang, 4:561

Mattaëlan Tree, 2:281

Mattahu malamaho, 3:34

Matthiola, 2:309*n8*, 5:328*n21*
  incana, 2:313*n4*

Mattua massin, 3:273

Maulanit, 4:470

Mauz, 4:6

Mayer, 4:159

Mazatli, 4:155

Mecapatli, 4:500

Medinilla
  crispata, 3:579*n1*, 5:540*on1*
  macrocarpa, 3:579*n6*

Meelappels, 2:428

Meha hatole, 5:71

Mehu
  huae, 3:315, 3:527
  mamulin, 5:60

Mehuaul, 5:95

Mehulappia, 3:401, 3:402, 3:421

Mehulatu, 4:521

Mela insana, 4:171*n15*

Melaleuca
  cajuputi, 2:103*n2*, 2:109*n1*,
    2:259*n29*, 2:279*n7*, 2:543*n2*,
    3:119*n5*, 3:149*n6*, 3:163*n3*,
    3:421*n5*, 3:520*n13*, 4:91*n7*,
    4:245*n10*, 4:321*n22*, 5:159*n7*,
    5:189*n19*, 5:221*n9*
  leucadendra, 2:102*n13*, 2:103*n2*,
    2:107*n15*, 2:109*n1*

Melanolepis multiglandulosa,
   1:279*n33,* 3:81*n3,* 3:377*n3,*
   3:407*n1,* 4:321*n24*

Melastoma, 3:426*n8*
   polyanthum, 3:427*n1,* 4:91

Melastomataceae, 3:401*n4,* 3:402*n9,*
   3:403*n1*

Meld-hout, 2:370

Melega, 4:105

Melica, 4:105

Melicera, 2:360

Melilotum, 3:255, 3:257

Melissa, 4:117*n2,* 4:207*n14,*
   4:251*n1,* 4:254, 4:255–4:256
   litorea, 3:308, 4:133, 4:251, 4:255
   officinalis, 4:257*n5*

Melithaea ochracea, 5:389*n38,*
   5:421*n1,* 5:421*n3*

Melium saburrum, 4:105

Melk-hout, 2:363

Melo
   aqueus, 4:431
   pepo indicus, *4:430,* 4:431–4:434

Melodinus, 3:494*n9*

Melodorum latifolium, 2:300*n6*

Melon, *4:430,* 4:435–4:436

Melongena, 4:171

Melostrobilus, 4:153

Memory plant, 5:52, 5:55

Menahu Cuhu, 3:90

Menampa Tree, 2:209

Mencudu, 3:53

Meninjo, 1:477

Menjom bulu, 3:507

Menoa, 1:395

Mentha, 4:205*n2*
   aquatica, 5:292, 5:293
   arvensis, 4:211, 4:213*n1*
   crispa, 4:212, 4:213
   pulegium, 4:188*n34,* 4:213*n2*
   spicata, 4:469*n2,* 4:469*n5*

Menthastrum amboinicum, 5:292,
   5:293

Merbou, 2:440*n8*

Merian pulli, 4:510

Merope angulata, 2:162*n6*

Merremia
   mammosa, 4:373*n1*
   peltata, 3:537*n6,* 4:475*n1,*
      4:507*n2,* 4:569*n4*
   umbellata, 4:191*n5,* 4:479*n1,*
      5:17*n6*

Mesantha, 4:66*n8*

Mespilus, 5:328*n22*
   germanica, 1:271*n3*

Mesua ferrea, 2:323*n18,* 3:591*n14,*
   5:469*n1*

Metrosideros, 2:569*n2*
   amboinica, 2:429, 2:438, 2:439–
      2:442
   macassarensis, 2:434, 2:435, 2:437
   molucca, 2:429, 2:443, *2:444,*
      2:445, 2:446
   spuria, 2:429, 2:447, *2:448,* 2:449
   vera, 1:432*n15,* 2:293*n16,* 2:429,
      2:431–2:432, *2:433,* 2:434–
      2:435, 2:436, 2:437, 2:505*n13,*
      2:581*n2,* 3:119*n8,* 5:563*n2*

Metroxylon, 3:49*n9*
   inerme, 1:298*n22*
   longispinum, 1:298*n21*
   micracanthum, 1:296*n17*
   sagu, 1:293*n1,* 1:296*n15,*
      1:296*n17,* 1:298*n20,* 1:298*n21*
   sylvestre, 1:298*n20*

Metroxylon sagu, 4:303*n8*

Mety, 3:254

Meum, 4:220*n13*

Meutan, 3:220

Meylantschi, 3:254

Meyr-boonen, 3:476

Miani puti, 4:251, *4:252,* 4:253

Mica mican, 4:217

Micca micca utan, 5:145

Michelia, 2:305
   alba, 2:303*n7*
   caerulea, 2:303*n6*

champaca, 2:55*n18*, 2:61*n2*,
  2:218*n4*, 2:299*n2*, 2:301*n1*,
  2:303*n5*, 2:303*n6*, 2:303*n10*,
  2:451*n4*, 3:8*n38*, 3:109*n2*,
  3:134*n4*, 3:427*n11*, 3:559*n12*,
  4:62*n27*, 5:196*n18*
parvifloria, 2:303*n5*
Michihoautli, 4:160
Micrococca mercurialis, 4:133*n2*,
  5:91*n1*
Microphyllos, 3:24
Micropiper silvestre fatuum, 3:550
Miga miga, 4:217
Milium
  Indicum, 4:103
  solis amboinicum, 5:40, 5:41
Milk Flower, 3:327*n3*
Milk Tree, 2:358, 2:359–2:361,
  2:362, 2:363, 2:396, 2:397–2:404
Milk-Giving Tree, 2:368, 2:369–
  2:371
Milkweed, 5:101*n2*
Milkwood, 1:327, 2:368, 2:369–
  2:371, 5:432*n16*
Milkwort, 4:146*n22*
Millefolium aquaticum, 5:330, 5:331
Millepora tenera, 5:435*n1*
Millet, 4:107, 4:115, 4:299,
  4:489*n3*, 5:7*n28*, 5:22
Millettia, 3:103*n1*
Milu milu, 4:79
Mimosa
  asperata, 4:270*n35*
  casta, 4:271*n43*
  dormiens, 4:270*n39*
  humilis, 4:271*n42*
  pudibunda, 4:271*n47*
  pudica, 4:270*n28*
Mimusops, 3:419*n1*
  elengi, 2:287*n1*, 2:289*n7*,
    2:303*n10*, 2:410*n2*, 2:434*n19*,
    2:549*n3*, 2:607*n3*, 2:625*n2*,
    4:453*n3*, 5:195*n5*, 5:197*n9*
  kauki, 2:434*n18*, 2:557, 3:107*n7*,
    5:468*n1*

parvifolia, 2:291*n1*, 2:394*n27*,
  2:499*n2*, 2:625*n3*, 4:62*n34*
Minibatu, 4:528
Minjassal unting, 5:9
Mint, 4:205*n2*
Mintäossa, 2:374
Mippi, 5:143
Mirabilis
  jalapa, 2:343*n3*, 3:223*n4*,
    3:230*n9*, 3:372*n15*, 4:191*n1*,
    5:115*n4*
  mervilje, 2:343
  mexicana, 2:343, 3:223, 3:372,
    4:192
  Peruana, 4:193
Miracle tree
  red, 3:341, *3:342*, 3:343
  white, 3:335–3:340
Mirobalane, 3:107
Mirobolanus emblicus, 2:343, 5:465,
  5:466, 5:467
Miscanthus, 5:29*n2*, 5:29*n3*, 5:29*n6*
  floridulus, 3:211*n1*
Mischocarpus fuscescens, 2:593*n6*
Mischophloeus vestiaria, 1:244*n29*
Miserly Leaf, 3:402
Missang midji, *4:2*
Mistballs, 5:251
Mistletoe, 1:224*n8*, 1:224*n9*,
  4:583*n4*, 5:173*n4*, 5:179, 5:180,
  5:184*n68*, 5:229
  Ambonese, 3:569, *3:570*, 3:571–
    3:572, 5:182*n49*
Mitra polonica, 5:443, 5:444
Moá, 4:119
Moatick, 3:188, 3:189
Mockulhäulo, 5:483
Modira caniram, 2:182
Modjo, 3:96
Moernagel, 3:371, 3:521, 5:439–
  5:440
Mohutoo, 3:109

Mokal, 3:381
  ahua, 3:385
Molineria capitulata, 5:215*n17*
Moluccan Oak Tree, 2:565, *2:566*, 2:567
Moluccan Seed, *3:344*, 3:345–3:347
Momentay wood, 3:65*n4*
Momordica, 4:447, 4:449
  charantia, 2:406*n11*, 3:417*n4*, 4:443*n2*, 4:447*n1*, 4:451*n1*, 4:531*n7*, 4:531*n8*, 5:71*n7*
  cochinchinensis, 4:453*n1*, 4:453*n2*
Moncus, 5:589–5:591
Money wort, 4:547, *4:548*, 4:549, *4:550*, 4:551–4:552
Mongo root, 5:521, *5:522*, 5:523–5:524
Monkey Cans, 3:675
Monochoria vaginalis, 5:335*n1*, 5:335*n5*
Monoclonos minor, 5:135
Montipora foliosa, 5:437*n3*
Moon leaf, 3:533, *3:534*, 3:535
  small, *3:536*, 3:537
Moon tree, 1:495*n4*
Moöputi, 3:16
Mopsella auruntia, 5:421*n4*
Moral, 3:382, 3:385
  babulu, *3:384*, 3:385
  catjangs, 3:385
Morangue gasts, 1:485
More more, 4:519
Moretiana, *5:100*, 5:101
Moretum malaicum, 4:523
Morfalla Tree, 2:481, 2:487, *2:488*
Morgenster, 2:468
Morian, 1:485
Moringa, 1:485
  bracteata, 3:51*n2*, 3:379*n20*, 4:138*n26*, 5:95*n7*
  citrifolia, 2:79*n2*, 2:193*n18*, 2:605*n7*, 3:53*n1*, 3:271*n20*
  eleifera, 4:63*n45*

  oleifera, 1:483*n1*, 2:15*n26*, 2:454*n8*, 2:497*n17*, 3:135*n5*, 3:333*n27*, 4:71*n1*, 4:326*n9*, 5:282*n17*, 5:465*n2*
Mort au chien, 5:310
Mortification-Root, 3:217*n2*
Morus
  alba, 5:479*n1*
  indica, *5:478*, 5:479
Mose, 5:249
Moss, 1:214, 3:91*n3*
  club, 5:165*n3*
  coral, 5:341
  hairy, *5:164*, 5:169–5:171
  Japanese yellow, *5:164*
  marine, 5:337*n1*
  shrub-like, *5:162*, 5:163
Mossel vrucht, 2:280
Mothe, 5:242, 5:243
Moti, 1:39
Mountain Amber, 2:237
Mountain Limis shrub
  rope-like, 3:510
Mountain Wabo Tree, 2:280
Mouse ears, 5:285, 5:346
Mouse Wood, *2:494*, 2:495–2:497
Moxa, 4:204
Mucuna
  aterrima, 4:395*n1*, 4:417In1
  gigantea, 3:481*n2*
  miniata, 3:483*n5*
  pruriens, 3:479*n2*, 4:419*n2*
  utilis, 4:395*n1*
Mudu, 1:443
Muebuae, 3:109, 3:113
Mugwort, *4:198*, 4:203–4:204
Mukia javanica, 4:537*n2*
Mukuy, 5:267
Mulberry tree, *5:478*, 5:479
Mulet tayn, 5:509
Mundholz-leim, 3:253
Mundubi, 4:474
Munniks keulen, 4:506

Murete plant, *5:100*, 5:101

Muri muri, 5:277

Murraya paniculata, 3:511*n1*,
  3:516*n18,* 3:560*n21,* 5:537*n1*

Musa, 4:2, 4:3–4:7, 4:9–4:12,
  4:13–4:17
  acuminata Colla, 4:23*n2*
  alphurica, 4:9, 4:21, 4:22
  brachycarpa, 4:11*n27*
  mensaria, 4:10
  paradisiaca, 3:45*n6,* 4:3*n1,*
    4:19*n7,* 4:445*n11*
    sapientum, 4:21*n2*
    troglodytarum, 4:19*n7*
  regia, 4:11
  sativa, 4:9
  silvestris, 4:9, 4:25
  simiarum, 4:23
  textilis, 4:25*n1,* 4:26*n17*
    amboinensis, 4:25*n4*
  uranoscopos, 4:19

Muscus, 5:169*n3*
  alga, 5:337
  capillaris, 5:170
  circinnalis, 5:170
  fruticescens
    foemina, 5:162, 5:163
    mas, 5:162, 5:163
  marinus, 5:337

Mushroom, 5:233–5:235, 5:444*n9*
  devil's yard, *5:240*
  earth, 5:243*n14*
  edible, *5:240,* 5:241–5:243
  funnel-shaped, *5:240*
  mouse-ear, *5:240*
  nutmeg, 5:239

Musk
  mallow, 1:432*n14*
  seed, *3:242,* 3:243–3:244
  wood, 2:63

Muskus-korrels, 3:243

Mussaenda
  forsteniana, 3:371*n7*
  reinwardtiana, 3:371*n1*

Mussel-Shaped Fruit Tree, 2:*278,*
  2:279–2:280

Mussi, 4:219, 4:220

Mussu, 3:43

Mustard, 5:220
  Chinese, 4:235–4:236

Mustard shrub
  wild, *3:592,* 3:593

Musus capillaris, 5:341

Muti
  salah, 5:425*n1*
  tanah, 5:425*n1*

Mutulabatta, 5:373, 5:425–5:427

Myagrum, 4:121*n13*

Mycrophyllos, 2:47

Myracanthus, 5:313

Myriopathes myriophylla, 5:381*n1*

Myristica
  fatua, 2:37*n1,* 2:101*n2*
  fatus, 3:603*n6*
  fragrans, 2:23*n1,* 2:26*n8,* 2:31*n12*
  odorata, 2:27
  radja, 2:26*n11*

Myrmecodia rumphii, 5:229*n3*

Myrrh, 2:59

Myrtle, 2:103, 2:111, 2:112, 2:113,
  2:575, 3:143, 3:265, 3:267
  pseudo Indian, 2:103

Myrtus
  Amboinica, 2:111, 2:112, 2:113
  angustifolia, 2:111
  communis, 2:111*n2*

N

Nachuni, 4:115

Naga
  pu, 4:248
  sari, 2:325, 5:469, *5:470,* 5:471
  valli, 3:469

Nago, 1:428

Nai corana, 4:420

Nail Flower, 3:251

Nail Wood, 2:589, 2:*590,* 2:591

Nala
  Hatu, 2:319
  kiri, 2:315
  Pata, 2:319
  tirtava, 4:207
  Uppar, 2:315

Nalla mulla, 3:560

Nalugu, 3:353

Nalur nalul, *5:40*, 5:43–5:45

Nam nam tree, *1:444*, 1:445–1:448
  wild, 1:449, *1:450*

Nana brawa, 4:156

Nanari, 2:220, 2:228, 2:231
  Minjac, 2:245

Nanarium
  minimum, 2:85, 2:236, 2:243,
    2:244, 2:245–2:247
  oleosum, 2:243, *2:244*, 2:245–
    2:247

Nanas utan, 4:225

Nanca, 1:345, 1:347
  bilulang, 1:344, 1:345
  bubur, 1:344, 1:345

Nanchauwa, 4:219

Nandi ervatam, 3:327

Nandu, 4:189

Nangka, 1:343*n1*

Nani
  hua, 5:468
  macassar, 2:557
  merak, 2:431
  wajer, 2:431

Nanium calapparium, 5:563

Nanschera canshabu, 4:535

Nantahule, 2:385

Nantsjera patsja, 4:545

Naragil, 1:196

Naransa Tree, 2:168

Narcissus, 1:335

Narcol, 1:196

Nardus stricta, 5:33*n4*

Nargel, 1:196

Nari patsja, *5:64*, 5:109

Narretali, 3:351

Nasturtium
  aquaticum, 4:525
  indicum, 4:235*n5*

Natsjoni, 5:18

Nauclea, 2:468*n11*

  mitragyna, 2:467*n4*, 2:563*n1*,
    2:563*n3*
  undulata, 2:467*n5*, 2:561*n1*

Naun, 1:273

Nawa, 1:273

Needle
  herb, 5:71
  lace, 4:528*n11*
  tree, 3:359

Negundo, 3:257, 3:258, 3:259

Nehulatu, 4:521

Nelen tsjindi

Neli, 4:110
  pouli, 1:424, 5:529

Nelumbium nelumbo, 2:262*n10*

Nelumbo nucifera, 3:77*n2*, 4:288*n24*,
    5:273*n19*, 5:319*n1*

Nenu Keri, 3:51

Neololeba atra, 3:671*n2*

Neonauclea, 2:427*n2*

Neonauclea moluccana, 2:483n1,
    2:483n3, 3:403*n2*, 5:499*n7*

Nepenthes, 4:123, 4:257
  maxima, 4:123*n1*
  mirabilis, 3:673*n1*, 4:123*n1*

Nepeta
  cataria, 4:257*n6*, 4:377*n3*
  sideritidis folio, 5:75

Nephelium lappaceum, 2:116*n6*

Nepu, 4:309

Nerium indicum, 5:495*n1*, 5:495*n6*

Nessaat Tree, *2:484*, 2:485–2:486

Nessatus, *2:484*, 2:485–2:486

Nettle, 3:217, 4:133, *4:254*, 4:255–
    4:256, 4:419*n4*, 5:57
  blind, 5:77*n3*, 5:279
  dead, 5:77*n3*, *5:88*, 5:91
  hemp, 5:73*n4*
  pig's, 5:90
  stinging, 5:77*n3*, *5:88*, 5:89–5:90
  white, 5:91

Neuburgia musculiformis, 2:279*n1*

New Leaf Tree, *2:330*, 2:331–2:334
  beach, *2:338*, 2:339–2:341

black, *2:342*, 2:343–2:346
creeping, 2:335
red, 2:337
Ngalea, 4:223
Ngaro, 3:91
Ngassi, 3:314
cototidi, 3:315
wild, 3:315
Nibum
mera, 1:242
kitsjil, 1:244
Nibun, 1:250
besaar, 1:242*n14*, 1:283
kitsyl, 1:283, 1:287
mera, 1:287
Nicolaia speciosa, 5:273*n15*
Nicotiana, 1:232*n11*
tabacum, 4:149*n1*
Niduli esculenti, *5:336*, 5:341–5:344
Nidus
germinans, *5:229*, *5:230*, *5:231*
vesparum marinus, 5:459
Nie tree, 1:281, *1:282*, 1:283–
1:284, 1:285, *1:286*, 1:287
red, 1:242
Nieboom, 1:283, 1:287
Nieuhof, Joan, 1:108, 1:119
Niewel-Latu, 1:203
Night Tree, *2:560*, 2:561–2:562
Nightshade, 4:171*n20*
black, 5:119*n1*
Niha Tree, 2:220
Niir Pongelion, 2:542
Nil, 4:143
Nila, 4:143
barudena, 4:171
Nila-hummatu, 4:182*n46*
Nilikay, 5:465, *5:466*, 5:467
Nimphaea minor, 5:335
Ninipattu, 1:362
Ninsin, 5:549
Nipa, 1:290
fruticans, 2:270*n18*, 3:435*n13*
Nippe, 1:290

Niri gohe, 5:459
Nisi, 5:549
Nitor Piperis Tree, 2:165
Nitu atoay, 3:215
Niu Aläa, 3:674
Nivuli, 3:333
Niwel
habu, 1:203
latu, 1:202*n25*
lupu, 1:203
yar issi, 1:202
Niwer, 1:196
Njalatta, 3:108
Njarang, 5:49, *5:50*, 5:51, *5:52*,
5:53
Njehio, 4:33
Njo, 1:196
Noli me tangere, 4:197, 4:269
Nopu, 5:301
Norchila, 3:257
Nos carpata, 3:336
Nothopanax, 2:80*n12*
Noti, 4:146
Notsja, 4:192
Nounouk, 3:9, 3:18
Noute, 5:153
Novella, *2:330*, 2:331–2:334, 3:220,
3:233, 3:243, 3:294
ebeniformix, *2:342*, 2:343–2:346
excellens, 2:335
litorea, *2:338*, 2:340–2:341, 3:131
nigra, *2:342*, 2:343–2:346, 2:412,
2:428, 2:437, 2:449, 2:495,
2:584, 3:71
repens, 2:335
rubra, 2:337
salimuri, 2:428, 2:437
secretorum, 2:335
Nuayhas, 3:194
Nucem medicam maldivensem, 5:390
Nucleos
pineos, 3:347
moluccos, 3:345, 3:346
Nude Tree, 2:574, 2:575

Nueburgia musculiformis, 2:279n6

Nugae
    silvarum, 3:631, *3:632,* 3:633
        miminae, 3:635–3:636

Nummularia, 1:207, 1:207n43
    lactea, 3:571, 4:553, 5:205
        major, 4:539, 4:547, *4:548,*
          4:549, 4:563, 5:513
        minor, 4:539, *4:550,* 4:551–
          4:552
    laganoides, 4:548, 4:553
    minima, 4:552

Nuno, *3:14,* 3:15–3:16, *3:17,* 3:18
    alulae, 3:21
    anahan, 3:5
    batali, 3:16
    hauessi, 3:16
        puti, 3:19
    laun maun, 3:24
    mattacau, 3:11

Nunuli, 3:295

Nurem kelangu, 4:363

Nussalavica tree, *5:490,* 5:493

Nussu, 1:466

Nutmeg, 1:39, 1:42, 1:43, 1:99,
    1:355, 1:357, 2:23, 2:24, 2:25–
    2:35, *2:36,* 2:37–2:39, *2:40,* 2:41,
    2:42, 2:43, *2:44,* 2:45, 2:46,
    3:121, 3:123, 3:215, 3:323, 3:571,
    3:637, 5:469, 5:473, 5:541, 5:566,
    5:567
    cultivation and harvesting,
        2:29–2:32
    male, *2:36,* 2:37–2:38
    mountain, 2:37
    uses of, 2:33–2:35
    wild, 2:39, *2:40,* 2:41, *2:42,* 2:43,
        2:44, 2:45, 2:46

Nux
    aromatica, 2:28
    faufel, 1:225
    indica, 1:196, 2:38
    metella, 4:179
    moschata, 2:28, 2:46
    myristica, 2:28, 2:37, 2:38
    vomica, 3:554, 4:179, 4:179n17

Nya, 2:220
    fanaâ, 3:425

Nycanthes, 2:303n10, 3:561
    arbor-tristis, 2:287n6, 2:289n7,
        3:559n11

Nyctaginaceae, 3:223n4

Nyit, 5:122

Nymphaea, 4:285, 4:291
    buronica, 5:326
    ceramica, 5:326
    indica minor, 5:325–5:329
    lotus, 5:325n2
    major, 3:77, 5:319, 5:320, 5:321–
        5:324
    nelumbo, 4:288n30
    pubescens, 4:36n20, 4:288n31,
        5:325n1, 5:326n7, 5:335n2
    stellata, 5:325n2

Nyor, 1:196

Nypa, 1:288, 1:289–1:291
    fruticans, 1:241n7, 1:289n1,
        2:427n6, 5:397n15, 5:520n8

NZambo cuneng, 1:404

O

Oats, wild, 5:26, 5:27

Obovy otoni, 3:234

Ochrocarpos ovalifolius, 2:367n13,
    2:589n1

Ochrosia oppositifolia, 2:370n4,
    2:385n20, 4:182n39

Ocimum
    basilicum, 4:205n3, 4:205n7,
        4:205n8, 4:205n9, 4:211n1,
        5:293n2
    indicum, 4:205, 4:206, 4:207
    sanctum, 4:205n6, 4:209n1
    tenuiflorum, 4:513n14

Ockir, *3:120*

Octomeles sumatrana, 3:127n1

Oculus
    astaci, 4:510, 4:561, *4:562*
    calidus, 5:51

Oculus Horti, 2:300

Oha lawa, 5:126

Ohulawan Aycou, 2:21, 2:22

Oimoá, 4:129

Ola-Ola, 3:155, *3:156*
Old Nick's Sari, 3:546
Oleaceae, 2:289*n7*, 2:451*n4*
Oleander, 3:286, 5:517*n2*
  sinicus, 5:*494*, 5:*495*
Oleum
  cicinum, 3:339
  cyprinum, 3:253
  fragrans, 3:254*n41*
  sambacinum, 3:559
Oli Oli, 3:476
Oloulo, 3:154
Olus
  album, 1:495, 3:154
    insulare, 1:497
  arundinaceum minus, 5:*40*,
    5:43–5:45
  calappoides, 1:313
  caprinum, 5:103
  catappanicum, 3:101
  Catholicum, 4:159
  crepitans, 4:544, 4:545, *4:546*
    mas, 4:*562*, 4:563–4:564
  crudum, 3:596, 3:597–3:598
  dulce, 3:350, 5:298
  javanum, 3:595
  litorale, 4:544
  marinum, 5:170
  mericum, 4:470
  nigrum, 4:237
  palustre, 5:335
  ranarum, 5:*330*, 5:331
  sanguinis, 4:568, 4:569
  scrofarum, 5:*64*, 5:65–5:66
  squillarum, 5:*68*, 5:69
  vagum, 4:461, *4:462*, 4:463,
    5:335
Ondo casturi, 4:359
Ongelboom, 5:476
Onglay, 2:585, 4:153
Onion
  Pekin, 4:83
Onion, wild, 5:*306*, 5:307–5:308
Onne Onne, 3:125
Oostindische, 5:108
Oostindise hulst, 5:311

Operculina turpethum, 4:373*n1*,
  4:477*n10*
Ophiocolla altera, 5:63, 5:99
Ophioglossum, 5:147*n2*, 5:157*n3*
  indicum simplex, 5:292, 5:295
  laciniatum, 5:292, 5:297–5:298
  pedunculosum, 5:295*n1*
  pendulum, 5:157*n1*
  reticulatum, 5:295*n1*
Ophioxylon serpentinum, 5:524*n18*
Ophyocolla, 5:60
Oplismenus compositus, 5:25*n2*
Opopanax, 2:80*n13*
  chironium, 3:303*n5*
Opossy, 1:244
Oppa, 4:354
Opuntia, 3:331*n13*
  tuna, 4:7*n45*
Orange apple, 3:169, 3:433, 3:557,
  3:616, 3:661
Orange flower tree, 2:*301*, 2:*302*,
  2:*303*–2:305
Orange tree, 2:*136*, 2:137
Orange-Apple tree
  Chinese, 2:*166*, 2:167–2:168
  sour, 2:163, 2:*164*, 2:165
  sweet, 2:169, 2:*170*, 2:171
Orchid, 3:8*n35*, 3:31, 3:571*n17*,
  3:673*n4*, 4:33, 4:579, 5:*178*,
  5:179–5:185, 5:223*n2. See also*
  Angraecum; Angrek
  bare, 5:207*n4*
  cassomba, 5:207*n6*
  champaca, 5:196*n21*
  coconut, 5:181*n33*
  downy, 5:206*n31*
  dusky, 5:197*n1*
  earth, 4:245*n8*
  Great White, 5:187*n8*
  ground, 5:217*n6*
  jambu, 5:207*n5*
  lemon, 5:197*n12*
  Mexican, 5:203
  narrow-leaved, 5:199*n13*
  narrow-purse, 5:201, 5:203*n15*
  purple, 5:215*n14*

purple forest, 5:*208*n8
  red, 5:191*n5*
  sinewy, 5:201*n6*
  small, white, 5:197*n2*
  terrestrial, 5:217*n5*
  tiger, 5:184*n59*
  white, 5:215*n14*
  wooly, 5:206*n31*
Orchis, 1:496*n6*
Orchis, 4:243, *4:244*, 4:245
  amboinica
    major, 5:223, 5:224, 5:225
    minor, 5:224, 5:227–5:228
  mascula, 5:213*n2*
  serapias, 4:243
Oregano, 5:555, 5:555*n2*
Oreocnide rubescens, 3:423*n1*
Oriental cotton tree, 1:503
Ormocarpum
  acutifolium, 3:137*n1*
  orientale, 2:495*n4*, 3:137*n1*,
    3:542*n15*, 5:475*n4*
  sennoides, 5:475*n4*
Ormosia calavensis, 3:85*n1*
Orta, Garcia da, 1:107, 1:108,
  1:111, 1:112, 1:115, 1:117
Ortelaen Ceylon, 4:213
Oryza, 4:110*n21*
  glutinosa, 4:113
  sativa, 4:107*n1*, 4:113*n1*
Osinibo, 5:273
Osmoxylon
  palmatum, 4:263*n5*
  umbelliferum, 1:454*n11*, 2:79*n1*,
    2:80*n14*
Osmunda, 1:317, 5:303
  arborescens, 1:311, *1:312*, 1:313–
    1:314, *1:315*, *1:316*, 1:317,
    1:318, 1:319–320
  cinnamomea, 5:125*n2*
  regalis, 5:125*n3*
Ossifraga lacteal, 5:*580*, 5:581–5:582
Ostindise, 5:311
Otanthera cyanoides, 2:485*n3*,
  3:425*n1*, 3:431*n2*
Ou, 4:287

Ou haytsjiu, 5:376
Oubak, 2:412
Oulasso vel macuda gangoli, 3:487
Outik, 3:206
Ouw, 1:331
Ova piscium, 4:98, 4:99–4:100
Ovifera, 3:36
Oxalis, 4:229
Oxalis repens, 1:31, 4:229In2
Oxtongue
  beach, *3:380*, 3:381–3:383
  wooly, *3:384*, 3:385
Oxyacantham, 3:254
  javana, 5:539
Oxycedrus, 2:145
Oxys lutea indica, 4:229, *4:230*
Ozfor, 4:136
Ozimum cirtatum indicum, 4:211,
  4:212

P

Paauw-staarten, 3:267
Pabulum cameli, 4:86
Pachyma
  cocos, 5:234*n15*, 5:235*n27*
  hoelen, 4:497*n7*, 5:234*n15*
  tuber regium, 5:233*n4*
Pachyrhizus erosus, 4:379*n1*
Pacolang, 5:273
Pacovos, 4:9
Pacu
  ager, 5:126
  alus, 5:140
  api, 5:131
  besaar, 5:126
  dibatan, 5:122
  kitsjil, 5:140
  lacki lacki, *5:128*, 5:131
  mera, 5:136
  pohon, 5:122
  rambat, 5:143
  tiang, 5:122
  udang, 5:136
    besaar, 5:133
  utan, 5:122

Pada kanka, 5:273

Padang
bilulangh, *5:16*, 5:17–5:18
buit buit, 5:22
jerac, 5:25, *5:26*
pangiris, *5:32*, *5:36*, 5:37–5:39

Padhac, 4:129

Padi
biji, 4:109*n15*
dalam, 4:109In14
ringan, 4:109*n13*
tahum, 4:109*n16*

Pady, 4:110
djidji, 4:109
taun, 4:109

Paederia foetida, 4:489*n1*, 5:77*n5*,
5:109*n18*

Paeonia
lactiflora, 5:552*n49*
moutan, 3:220*n29*, 5:552*n52*

Paerou, 4:403

Pagamatta tree, *3:64*, 3:65

Pagoda tree, 3:27*n6*

Pagode-boom, 3:27, 3:30

Pahaytsjiu, 5:376

Pahudia galedupa, 2:85*n1*

Paint Tree, *2:114*, 2:115–2:117

Pajomirioba, 5:283

Pajong, 5:242
laut, 2:639

Pakis, 5:129
bitbit, 5:135
catoc, 5:126
radji, 1:313
ticko, 5:135

Paku duke, 5:151

Pala
Bacambar, 2:26
Domine, 2:26
Fuker, 2:37
Hollanda, 2:26
Kackerlack, 2:26
Lacki-lacki, *2:36*, 2:37–2:38
Pentsjuri, 2:26
Puti, 2:26

Radja, 2:26
Utan, 2:37

Palacca tree, 3:127, *3:128*, 3:129

Palala, 2:37, 3:603, 3:637
ala Manay, 2:39
aruana, 5:567
Burong, 2:39, 2:41
Canari, 2:43
Dauer Kitsjil, 2:43
dentaria, 2:45
Laun Mâun, 2:43
minima, 4:175
niggicau, 4:175
quarta, 2:43
quinta, 2:43, 2:45, 2:46
secunda, 2:39, *2:40*, 2:41, 2:43,
2:46
sexta, 2:46
tertia, 2:43
tingens, 2:43
Utan, 2:39

Palan, 4:6

Palaquium, 2:467*n7*, 2:529*n2*
amboinense, 2:475*n1*
macrocarpum, 2:475*n4*
obovatum orientale, 2:475*n4*

Palatsgayt, 3:336

Paleam de meckay, 4:86

Palean, 1:290

Paleca, 1:226

Palembang, 1:34

Palijuncus calapparius, 3:639, 3:640

Paliurus, 1:506

Palm
aren, 1:340
cabbage, 3:305, 3:305*n4*,
3:441*n4*, 3:449
climbing, 1:334*n10*
coconut, 4:25*n2*, 4:123*n3*,
4:137*n22*, 4:267*n3*
date, 1:298
dwarf, 1:197*n71*
gomuti, 4:25*n3*
Great, 1:201*n3*
Indian, 1:189–1:199, *1:200*,
1:201–1:219, 1:220, 1:221,
1:222, 3:640, 4:4

mealy, *1:292, 1:293–1:296,*
  1:297, 1:298–1:300
nipa, 1:242*n16*
pinang, 4:125*n14*
reed, 3:637, *3:638,* 3:639–3:642
  black, 3:643, *3:644*
  horseman's, 3:657, *3:658,*
    *3:659–*3:670
  smooth, 3:671, *3:672*
  true, 3:649, *3:650,* 3:651–
    3:653
  white, 3:645, *3:646,* 3:647–
    3:648
  wicker, *3:654,* 3:655–3:656
sago, 1:1, 1:242*n16,* 4:78*n5,*
  4:303*n8,* 5:282*n18*
Small, 1:201*n4*
sugar, 4:287*n16*

Palma
  Brasiliana, 1:284
  Christi, 3:336
  coccifera, 1:261
  cretica spinosa, 1:298
  cypria, 4:7
  d'Igresia, 1:319, 1:320
  dactylifera, 1:284
  espine, 1:298
  filix, 5:303
  indica, 5:121
  Indica
    major, 1:189 1:199, *1:200,*
      1:201–1:219, *1:220,* 1:221,
      1:222
    Montana, 1:249
    tal, 1:261
    talghala, 1:261
  marina theophrasti, 5:376
  minorem, 1:201
  nucem ferentem, 1:201
  vinaria secunda, 1:273
Palmeer-boom, 1:256
Palmeira, 1:196
  Brava, 1:255, 1:256
Palmifilix
  alba, 5:121–5:123
  nigra, 5:121–5:123
  postium, 5:121–5:123
Palmijuncus, 3:637, *3:638,* 3:639–
  3:642

albus, 3:639, 5:408
anguinus, 5:369, 5:371
aracanicus, 3:652, 3:653
draco, 3:663–3:664, *3:665,* 3:666–
  3:669, 5:540
equestris, 3:639, 3:657, *3:658,*
  3:659–3:670
fimbriarum, 3:652
graminosus, 3:645, *3:646,* 3:647–
  3:648
laevis, 3:671, *3:672*
marinus, 5:369, 5:370, 5:371–
  5:373
  albus, 5:409–5:410
niger, 3:639, 3:643, *3:644*
palimbanicus, 3:653
striatus, 5:369, 5:371
unicaulis, 3:649, *3:650,* 3:651–
  3:653
verus, 3:639, 3:649, *3:650,* 3:651–
  3:653
  angustifolius, 3:651, 3:652
  latifolius, 3:651
viminalis, 3:639, *3:654,* 3:655–
  3:656
vulgaris, 5:369
Palmitum, 1:197
Palmum majorem, 4:4
Palmyra tree, 1:256
Palmyruncus Callapparius, 2:383
Palun, 1:287, 3:271
Palunen parun, 1:283
Pamali, 2:473
Pana, 5:273
Pana-Docki, 2:406
Panagay, 1:256
Panamaran, 1:256
Panas, 1:344
Panava, 3:347
  xula, 3:355
Panawa, 4:445
  massou, 2:192
  pipis, 2:192
  sassiri-isso, 5:223
  xula, 2:511
Panax ginseng, 5:549*n30*

Pancaga, 4:513, 4:523, *4:524*, 4:525, 4:527

Pancaluhur, 4:165*n3*

Pancratium zeylanicum, 5:205*n23*, 5:301*n13*, 5:309*n1*

Pandanas spurius, 3:437, *3:438*, 3:439

Pandang, 1:208, 4:109, 4:299, 5:393
  ayer, 3:463
  babouw, 3:447
  bastard, 3:437, *3:438*, 3:439
  boubou, 2:381
  broad-leaved, *3:446*, 3:447
  casturi, 3:449–3:450
  ceram, 3:451, *3:452*
  creeping, 3:459
  gunong, 3:443
  kitsjil, *3:440*, 3:441–3:442
  land, 1:242*n13*, 3:443*n7*
  lengis, 3:450
  low, *3:440*, 3:441–3:442
  matti, 3:437, *3:438*, 3:439
  mera, 3:461
  musk, 3:449–3:450
  nipa, 3:435
  rope-like, *3:460*, 3:461
  tali, *3:460*, 3:461
  true, *3:432*, 3:433–3:436, 3:441, 3:447
  utan, 3:443, *3:444*, 3:445
  wild, 3:443, *3:444*, 3:445
  with alang-alang leaves, 3:463
  wong, 3:453

Pandang-lang, 1:242

Pandanus, *3:432*, 3:433–3:436
  amboinensis, 3:443*n4*, 3:445
  bagea, 1:290*n6*, 2:220*n16*
  caricosus, 3:461, 3:463, 5:38
  ceramicus, 3:451, *3:452*
  conoideus, 3:451*n1*
  dubius, 1:219*n4*, 3:455*n1*
  funicularis, *3:460*, 3:461
  hasskarlii, 2:381*n4*, 3:447*n1*, 4:287*n19*
  humilis, *3:440*, 3:441–3:442
  laevis, 3:449–3:450
  latifolius, *3:446*, 3:447
  littoreus, 4:11, 4:155, 4:156
  moschatus, 3:449–3:450

odoratus, *3:446*, 3:447
polycephalus, 3:441*n1*, 4:11*n30*, 4:155*n13*, 4:579*n5*, 5:304*n23*
repens, 3:453*n4*, 3:459, 5:95*n12*
robinsonii, 3:437*n1*
silvestris, 3:463
  sive terrestris, 3:443, *3:444*, 3:445
tectorius, 1:208*n51*, 1:242*n13*, 1:431*n11*, 1:505*n6*, 3:433*n7*, 3:449*n1*, 3:455, 4:99*n2*, 4:109*n17*, 4:151*n3*, 4:299*n4*, 5:392, 5:497*n4*
terrestris, 3:445, 4:155*n12*
verus, 3:435, 3:443

Pandanus bagea, 3:453*n1*

Pandi avanacu, 3:341

Pandorea, 5:543*n5*

Pandy pavel, 4:449

Pangaga, 2:349

Pangel, 5:507
  boaja, 5:507

Pangi, 2:274, 2:275–2:277, 3:127

Pangium, 2:274, 2:275–2:277
  edule, 2:275*n1*, 2:279*n6*, 2:349*n7*, 2:454*n7*, 3:127*n3*, 4:182*n37*

Pangran, 4:153

Panic plant, *4:105*, 4:115–4:116, *4:118*

Panick corn, 5:34

Panicum, 5:18, 5:591
  gramineum, 4:116
  Indicum, *4:105*, 4:115–4:116, *4:118*
  reptans, 5:22*n7*

Pannas, 5:269

Pannekoeken, 5:446

Pantsja loho, 4:165

Panu itam, 5:93

Panubessi, 5:93

Panulirus versicolor, 4:561*n6*

Pao de aquila, 2:54

Pao de Brasil, 3:272

Pao de chanco, 5:527

Pao de Cobra, 2:181

Pao de Sappan, 3:272

Pao Salgoedo Macho, 2:604

Papaja, 1:412, 1:413, 1:414–1:415,
1:416, 1:417–1:420, 1:421
beach, 1:419
femina, 1:417
littorea, 1:420
maram, 1:418
mas, 1:417
panate, 1:420
silvestris, 1:420
utan, 1:419, 1:420

Papaja tree, 5:477

Papari, 4:447
utan, 4:450, 4:451

Paparihan, 4:447

Papaver, 3:173
somnifrum, 5:325n3

Papaveraceae, 4:59n9

Papaya, 1:414–1:415, 1:417–1:420,
3:220n21
femina, 1:261
mas, 1:261

Papayta, 4:529

Papeda, 3:117

Papegaye-kruyd, 4:168

Paper Mulberry tree, 3:377n2

Paper shrub, 3:377

Papeyte albar, 5:136

Papie rope, broad-leaved, 3:491,
3:492, 3:493, 3:494–3:495

Pappa, 1:317

Pappatale, 2:332

Pappe Pauna Tree, 2:123

Paprihan, 4:451

Papua wood, 2:83–2:84

Paradise wood, 2:47–2:51

Paradys-hout, 2:49

Parameria barbata, 2:15n23,
3:517n1, 3:609n5

Parammuli, 3:295

Parampuan, 5:99, 5:364

Parang, sea bean, 3:474, 3:475–
3:478

Parangian, 3:300

Paranthipathes larix, 5:377n6

Parari peria, 2:406n11

Paras laut, 3:595

Pare, 4:110

Parena, 1:290

Parens muscarum, 5:494, 5:497

Paria laut, 3:595

Parietaria, 5:47n3, 5:83
amboinica, 5:52, 5:55

Parihan abbar, 4:451

Parinarium glaberrimum, 2:249n2,
2:275n3, 2:588n4, 4:300n12

Pariti, 2:337, 3:220, 3:239

Parkia speciosa, 2:496n12

Parkinsonia, 3:136, 3:137

Parococuli, 4:179

Parrab kitsjil, 3:481, 3:482, 3:483

Parrana
hutan, 3:477
mette, 3:481, 3:482, 3:483
miniata, 3:483
nigra, 3:476
rubra, 3:479, 3:480

Parrang, 4:387

Parrot's plant, 4:168n15

Parsley, 5:115, 5:318n18
Javanese, 4:219
sea, 5:315n2, 5:315n5
wild, 5:23n3

Parsnip, water, 3:313n3, 4:523,
4:523n7, 4:525

Partago, 5:16, 5:17–5:18

Parthenium sinense, 4:198, 4:199–
4:201

Parun maun, 1:287

Paspassan, 4:511, 4:512, 4:513,
4:569

Passal
Hatu, 2:452
Mahina, 2:452
Upal, 2:452

Passee, 4:186

Passiflora, 3:247*n5*

Passion Flower, 3:247

Passo baguala, 3:635

Pastinaca sativa, 4:219*n3*

Patchouli, 4:255*n2*

Pate abbal, 3:413

Paternoster plant, 4:79

Patiatta Tree, 2:293

Patienta, 4:229

Patsin cherra tadha, 5:174

Patsjing, 5:277

Patta
    camudi, 4:261
    tulang, 5:581–5:582

Pattene Tree, 2:315

Patuku, 1:313

Patul, 1:451

Pau, 1:329
    fule, 2:305

Pausengi, 5:384

Pausoldo, 3:79

Pauw tree, *5:500*, 5:501

Pawpaw, 3:220*n21*

Paya matta, *5:13*, 5:15

Payday tree, 3:78

Paydoh, 4:523

Payena leerii, 2:525*n1*, 3:143*n5*

Paynga, 1:225

Pea, 5:39*n20*
    green, 5:126*n10*

Peacock Tree, 2:571

Peacock's Crest, *3:264*, 3:265–
    3:267

Peamatta, 3:65

Pear, 3:18, 3:119
    balsam, 4:447*n1*
    Chinese, 1:395
    prickly, 4:7*n45*

Pearl shrub
    first, *3:386*, 3:387–3:388
    second, 3:389, *3:390*, 3:391

Pecanga, 4:500

Pectellis susannae, 1:64, 4:243*n1*

Pectinia lactuca, 5:437*n3*

Pedilanthus, 3:329

Pee
    candel, 2:604
    inota inodien, 5:117

Pegamatta, 2:522

Pegang Matta, 3:65

Pekin zoan, 4:83

Pelagium, 5:416

Pelal, 3:13

Pele tsjedangan, 3:555

Pellionia sinuata, 5:257*n1*

Pelola sylvestris, *4:444*, 4:445–4:446

Pemphis acidula, 2:411*n13*, 2:631*n1*

Pencil Cedar, 2:569*n2*

Penicillus marinus, 5:457

Pennyroyal, 4:188*n34*, 4:213*n2*

Pentapetes phoenicea, 4:247*n1*

Penthouse flower, 3:555

Peony, 3:220*n29*
    tree, 3:220*n29*

Pepe pepe, 2:623

Pepinjo ticos, 4:537

Pepo indicus, 4:427, 4:429

Peponaster, 4:555–4:556

Peponifer, 1:419

Pepper, 1:37, 1:41
    black, 4:316, 4:317
    Brasilian, 4:183, 4:189
    catjang-shaped, 4:316
    chili, 1:236*n43*
    hot, 5:131*n5*
    Java, 2:7*n15*
    leaf, 5:279*n6*
    long, 1:236*n42*, 4:314, 4:315–
        4:317
    of the Indies, 4:189
    siri-shaped, 4:315
    Spanish, 4:188
    West-Indian, 4:189
    white, 4:317

Pera, 1:405

Peralu, 3:36

Peregrinus fructus, 3:96

Perexil de mar, 5:315

Pericampylus glaucus, 3:537*n1*, 4:481*n10*

Periclymenum, 3:300

Perim-kaku valli, 3:478

Perinde coddy, 3:611

Periploca, 3:495, 3:495*n10*, 3:496, 3:497
   graeca, 4:541*n9*

Peristrophe bivalvis, 5:95*n1*

Peristylus goodyeroides, 5:227*n10*

Periwinkle, 4:419*n1*

Perlarius
   alter, 3:389, 3:390, 3:391
   latifolius, 3:388*n10*
   primus, 3:386, 3:387–3:388

Perrottetia moluccana, 2:597*n5*, 5:481*n3*

Perticaria, 4:157, 4:183*n12*
   ferria parvifolia, 2:555, 2:556, 2:557
   nigra, 3:113
   prima, 3:109, 3:115
   secunda, 3:113
   tertia, 3:115

Peru Teregam, *3:42*

Pervatti pou, 3:300

Pes
   elephanti, 4:477
   equinus, 4:523, *4:524*, 4:525
   locustae, 5:174

Pestilence Root
   Ambonese, *3:362*, 3:363–3:364
   wild, 3:365, *3:366*, 3:367

Petasites, 5:259, 5:485
   agrestis, 3:365, *3:366*, 3:367
   ambonica, *3:362*, 3:363–3:364

Petattes, 4:372

Peterceli do maar, 5:315

Petersely de Maer, 2:643

Petola, *5:172*, 5:175–5:176
   anguina, *4:440*, 4:441
   bengalensis, *4:442*, 4:443
   djingi, 4:439
   panjang, 4:437, *4:438*, 4:439
   risca, 4:443
   silvestris, 5:367
   snake-shaped, *4:440*, 4:441
   tchina, 4:437, *4:438*, 4:439
   ular, *4:440*, 4:441, 4:443
   utan, 4:439, *4:444*, 4:445–4:446, 4:446

Petraeovitex multiflora, 3:473*n1*

Petroselinum maritimum, 5:315

Petume, 4:150

Peucedanum ostruthium, 3:301*n4*

Peulen, 4:392

Peylale, 3:19

Phacelophrynium robinsonii, 4:29*n13*

Phaisus
   terrestris, 5:217*n1*

Phalaenopsis, 5:175*n7*
   amabilis, 5:187*n1*, 5:187*n7*
   deliciosa, 5:187*n6*

Phalaris canariensis, 5:18*n8*

Phallus
   daemonum, 3:177, 5:253
   impudicus, 5:235*n28*

Pharmacum
   legitimum, 2:201, 2:202, 2:203–2:204
   limonicum, 2:201, 2:202, 2:203–2:204
   litorcum, 3:617, *3:618*, 3:619
   magnum, 3:540, 3:541–3:543, 3:554
   nummularium, 2:192
   papetarium, 3:420, 3:421
   sagueri, 2:202, 3:151

Phaseolus
   adhaerens, 5:289
   aureus, 2:360*n5*, 3:103*n7*, 4:232*n12*, 4:405*n1*
   balicus, 4:386, 4:389
   calcaratus, 4:411*n1*
   cylindricus, 4:401*n1*, 4:411
   de Jemen, 3:567

Indian purging, 3:564
indicus
   minimus, 2:360
   ruber, 3:567
Javanus, *4:400*, 4:401–4:403
litoreus, 4:412, 4:415
marinus, 4:412, 4:415, 4:485
minimus, 4:405–4:407, 4:415
minor, *4:400*, 4:401–4:403,
   4:401*n1*
   ruber, 4:401
montanus, 5:280, 5:281–5:282
mungos, 4:405*n1*
niger, *4:408*, 4:409–4:410
perennis, 4:391–4:392
radiatus, 1:488*n16*, 3:35*n10*
ruber, 4:401
rumphii, 3:353*n9*, 3:382*n13*,
   3:385*n2*
scriptus, 4:399
sinensis, 4:383, *4:384*, 4:385
vulgaris, 4:379*n3*, 4:399*n1*,
   5:126*n10*
Phasius
   gratus, 5:215, 5:215*n11*
Phipeleg, 4:48
Phoebe, 2:532*n9*
Phoenix
   Amboinica montana, *5:32*, 5:35
   mountain, *5:32*, 5:35
Pholidocarpus ihur, 1:269*n1*
Phragmites
   karka, 5:29*n7*
   vallatoria, 3:209*n1*
Phul, 4:392
Phyllanthus, 5:80*n7*
   acidus, 1:424*n11*, 3:69*n2*,
      5:529*n1*, 5:529*n12*
   emblica, 1:409*n7*, 3:107*n5*,
      5:465*n1*, 5:465*n6*, 5:465*n8*
   niruri, 5:27*n8*, 5:66*n12*, 5:79*n3*
   urinaria, 5:53*n22*, 5:66*n12*,
      5:79*n6*
Phyllanthus emblica, 2:124*n15*,
   2:229*n3*, 2:343*n4*
Phyllitis
   amboinica, 5:153, *5:154*, 5:155
   arborea, 5:153, *5:154*, 5:155
   terrestris, 5:153, *5:154*, 5:155

Physalis
   alkekengi, 5:115*n3*
   angulata, 4:513*n11*
   minima, 3:368*n3*, 5:115*n6*
Phytographia, 5:170*n7*
Picanna, 4:439
Piccat, 3:477
Picria fel-terrae, 4:234*n9*
Pictiansom, 5:548
Pidjar, 3:567
Piessang rope-plant, *3:602*, 3:603
Pigafetta
   filaris, 3:647*n6*
   filifera, 3:453*n5*, 3:669*n22*
Pigmentaria, 2:114, 2:115–2:117
Pigweed, 5:545*n6*
Pilo carandi, 4:363
Pilosa, *5:554*, 5:555
Pilosella ambonica, 5:285
Pimela caryophyllacea, 2:228*n15*,
   5:468*n1*
Pimelodendron amboinicum, 2:377*n1*,
   2:443*n2*, 3:134*n7*, 3:641*n18*,
   4:182*n38*
Pimenta
   d'Brasil, 4:186, 4:189
   dos Indos, 4:189
Pimperling, wild, 4:517
Pimpinella
   anisum, 2:97*n18*
   saxifraga, 5:140*n13*
Pina hui huitzle, 4:271
Pinaceae, 2:569*n2*
Pinang, 1:223–1:239, 1:240, 1:241–
   1:242, 1:243, 1:244, 1:245, 1:246,
   1:247, 1:285, 1:313, 2:258, 3:119
   acorn-like, 1:241
   alva, 1:226
   bacambar, 1:228
   boubou, 1:227
   calapparia, 1:223*n1*, 1:226
   coconut, 1:226*n29*
   itam, 1:227, 1:229, 1:241
   kachi ayan, 1:227
   lanssa, 1:228, 1:242
   mabok, 1:231

mountain, 1:239
muda, 1:225, 1:226
nigra, 1:227
oryzae-formis, 1:246
pandang, 1:242
puän, 1:227
puti, 1:226, 1:227, 1:229, 1:373
rice-grain, 1:242, 1:244
saleê, 1:244
small, 1:244, 1:244n20
telloor, 1:227–1:228
touwa, 1:225
tsjelacatte, 1:226
utan
   besaar, 1:239, 1:239n5
   kitsjil, 1:244
white, 1:226
wild, 1:239, *1:240*, 1:241–1:242,
   1:243, 1:244, *1:245*, 1:246,
   1:250
wilde ronde, 1:239
zwarte, 1:227

Pinanga, 1:225, 3:213n8
globosa, 1:239
globulifera, 1:242n19
oryzae-formis, 1:242
punicea, 1:241n11, 1:287n8
sylvestris
   glandiformis, 1:241, 1:242
   globosa, 1:239
   oryzae-formis, 1:244

Pinas, 4:153

Pinea, 4:155

Pineapple, 1:311, 1:311n5, 1:313,
   1:314, 1:317, 1:341n39, 3:433n3,
   3:434, 3:435n11, 4:151, *4:152*,
   4:153–4:156, 4:547n3, 4:579n3

Pingir ayer, 3:617

Pinoguacu, 1:419

Piper
amboinense, 3:549n16
arborescens, 3:547n1, 3:551n2
attenuatum, 4:333n1
betle, 1:225n14, 1:231n3,
   1:232n11, 1:339n30, 2:115n3,
   2:473n3, 2:521n4, 2:521n7,
   3:47n2, 3:78n11, 3:153n2,
   3:321n2, 3:356n7, 3:461n4,

3:541n3, 3:660n14, 4:81n2,
   4:137n20, 4:173n6, 4:283n2,
   4:317n15, 4:319n1, 4:319n2,
   4:325n1, 4:325n2, 4:331n2,
   4:538n5, 4:561n3, 4:582n6,
   5:13n8, 5:39n17, 5:72n11,
   5:126n12, 5:157n3, 5:389n4
caducibracteum, 3:531n6,
   3:573n1, 4:329n1, 4:510n17
caninum, 2:89n5, 3:542n7, 3:551,
   3:551n1
cubeba, 2:89n5, 5:555n11
decumanum, 3:545n1, 3:545n6,
   3:551n2
fragile, 3:541n2, 4:515n6
gelalae, 4:329n2
indicum, 4:186
longum, *4:314*, 4:315–4:317,
   4:319n3
nigrum, 4:317n17, 5:346n18
reinwardtianum, 3:442n6,
   3:545n6
retrofactum, 4:315n2, 4:316n14,
   4:319n3, 4:515n6, 5:39n18
retrofructum, 1:236n42, 3:541n2,
   3:545n7
sarmentosum, 4:331n1
siroides, 4:315
subpeltatum, 5:259n1
sylvestre macro, 3:547, *3:548*,
   3:549–3:550

Piperinus, 5:482, 5:483–5:484

Piperitis indica, 5:278, 5:279

Pipinho, 4:435

Pipturus
argenteus, 3:387n1, 3:388n10,
   3:388n14
repandus, 3:581n13

Pisang, 1:204, 1:417, 1:419, 3:176

Pisonia
aculeata, 3:510n24
alba, 3:154n5
grandis, 1:57, 1:495n1, 1:497n1
   sylvestris, 5:193n7
sylvestris, 1:497n2

Pissang, 1:208, 4:149
abu, 4:11, 4:11n29
alfuran, 4:9

alphuru, 4:11, 4:21, *4:22*
Ambon, 4:10*n20*
batu, 4:11, 4:11*n27*, 4:15, 4:21, 4:25
Bengal, 4:153
berangan, 4:10*n19*
bidji, 4:11
biokaykeck, 4:27
bolvanger, 4:10*n18*
bombor, 4:11, 4:11*n30*
boratsjo, 4:10
bulang trang, 4:12
canaya, 4:23
    kitsjil, 4:12
    puti, 4:11
carbou, 4:10, 4:21
Ceramese, 4:21, *4:22*
croho, *4:2*, 4:10, 4:11, 4:14, 4:15
culit tabal, 4:10
cutsjupau, 4:11
djernang, 4:10, 4:23, 4:23*n1*
gaba gaba, 4:10
garden, 4:9
horned, 4:9–4:10
jacki, 4:23
kapas, 4:10*n10*
keker, 4:23
kepok, 4:11*n29*
key, 4:10
maas, 4:10
medji, 4:4, 4:4*n14*, 4:10, 4:11, 4:15, 4:16, 4:21, 4:23
merah, 4:11, 4:11*n22*
monkey, 4:12, *4:22*, 4:23
needle, 4:10
ocky ocky, 4:10
radja, 4:11, 4:11*n21*, 4:12
    sere, 4:12In31
salpicado, 4:11
shrub, *4:2*, 4:3–4:7
sky-pointing, 4:19
soldado, 4:11
sussu, 4:12
swackan, 4:10
swangi, *4:2*, 4:11, 4:15, 4:16, 4:21
table, 4:10*n20*
tame, 4:9
tando, 4:9–4:10, 4:10, 4:21
toncat langit, 4:19
tundjoc langit, 4:19

ubi, 4:10
utan, 4:25
warangan, 4:10, 4:10*n19*
wild, 4:9, 4:25–4:26
Pistacia lentiscus, 2:135*n20*, 2:265*n9*
Pistia stratiotes, 4:243*n4*, 5:333*n1*
Pitcher plant, 3:673–3:676, 4:123*n1*
Pithecellobium, 3:119*n9*
    clypearia, 3:87*n5*, 3:135*n6*
Pitorik, 1:463
Pitsja, 1:313, 3:403
Pitsjaar tree, 3:3*n2*, 3:26, 3:27, *3:28–3:29*, 3:30–3:31
Pitsjaar-boom, 3:27, 3:30
Pittis pittisan, 4:549, 4:551
Pittosporum, 5:483*n1*
    molucanum, 5:487*n1*
Pityriasis versicolor, 2:461*n4*
Plaintain, 4:3*n2*
Planta
    anatis, 4:226, 4:227–4:228
    bantimica, 5:108
    sentiens, 4:271
Plantago, 4:243*n4*, 5:221*n4*
    aquatica, 5:333–5:334
    major, 4:497*n2*
Plantain, 2:521, 4:497*n2*, 5:44*n16*, 5:221*n4*
    broad leaved, 4:497
    water, *5:330*
Plantam saxeam abrothanoidem, 5:430
Plantano, 4:3
Plantarum
    americanum, 4:502*n13*
    historiae universalis, 2:384*n18*
Platanaria, 5:38
Platpimperling, 4:516
Platter shrub
    first, 3:301, *3:302*, 3:303
    second, 3:305, *3:306–3:307*, 3:308
    third, 3:309, *3:310*, 3:311

Platycerium coronarium, 2:222*n28*

Plectranthus, 4:497*n3*, 5:77*n2*
   amboinicus, 4:257*n1*, 5:80*n13*
   scutellarioides, 4:69*n15*, 4:251*n3*,
      5:65*n3*
   tuberosus, 4:377*n1*

Plegmaria celonica, 5:174

Pleomele angustifolia, 2:384*n16*,
   3:315*n1*, 5:135*n7*

Plisse dangan, 3:555

Plotosus lineatus, 2:345*n15*, 5:539*n5*

Ploughman's spikenard, 5:107*n1*

Pluchea indica, 4:263*n6*

Pluin-keylen, 4:351

Plukenetia corniculata, 1:499*n9*

Plum tree, 3:145

Plumapathes pennacea, 5:381*n2*,
   5:381*n4*

Plumbago indica, 3:359*n7*, 4:521*n1*

Plumeria
   acuminata, 2:191*n2*
   rubra, 3:325*n1*, 3:325*n5*

Poa amboinica, 5:35*n1*

Poac, 1:226

Poangus, 1:325

Pockwood, 4:495*n19*

Pod-bearing tree, East-Indian, 3:84

Podocarpus rumphii, 5:503*n6*

Pogostemon
   auricularius, 5:77*n1*, 5:171*n24*
   cablin, 2:598*n8*, 4:133*n3*,
      4:255*n1*

Pohon kofo, 4:25*n6*

Poinciana, 3:267

Poison
   dart, 2:405–2:406
   tree, 2:396, 2:397–2:404,
      5:299*n2*

Pokhout, 4:495

Pokor Cupan Wassi, 2:311

Pokwood, 2:414, 2:431

Pola pola halauhe, 5:126

Polang laut, 5:439

Polanisia viscosa, 4:233*n2*, 4:523*n15*

Polat Puti, 3:176

Pole tree, 2:592, 2:593, 2:594,
   2:595, 3:109, 3:114, 3:115

Polemonium coeruleum, 5:184*n59*

Polendo darat, 3:381

Poleye, East-Indian, 4:212, 4:213

Polianthes tuberosa, 2:303*n10*,
   4:241*n1*, 5:223*n5*

Polishing Leaf, 3:404, 3:405–3:406

Pollutamel, 4:300

Polot, 2:81, 3:174, 3:175–3:177

Polyalthia, 2:300*n4*, 2:421*n1*,
   2:422*n5*, 2:424*n9*

Polygala ceylonensis, 4:146

Polygaster sampadarius, 5:237*n1*

Polygonatum, 5:297*n4*

Polygonum
   Amboinicum, 5:21–5:22
   aquaticum, 5:19

Polyphyllia talpina, 5:443*n4*

Polypodiaceae, 5:123*n17*, 5:125*n4*

Polypodium, 1:102, 1:271, 1:311,
   1:327, 3:24, 5:140, 5:146,
   5:153*n2*, 5:157
   albens, 5:133*n4*
   filix femina, 4:270*n30*
   indicum, 1:271*n1*
      glabrum, 5:151
      majus, 5:148, 5:149–5:152
      minus, 5:151
   phymatodes, 5:151*n19*
   pilosum, 5:148, 5:149–5:152
   sinuosum, 5:157*n5*

Polyporus, 5:245*n1*, 5:245*n4*

Polyscias
   fruticosum, 3:309*n1*, 5:346*n12*
   nodosa, 1:419*n19*, 5:477*n1*
   pinnatum, 2:593*n5*
   scutellaria, 3:301*n1*, 3:305*n3*,
      3:349*n2*
   tricochleata, 3:305*n1*

Polystictus sanguineus, 5:245*n5*

Polytrichum
  commune, 5:31*n2*
  gramen, 5:31, i5:32
Poma amaris, 5:115
Pomegranate, 1:391*n13*, 2:136,
  2:137–2:138, 3:249, 3:251, 3:513
Pomelo, 1:325*n12*, 1:341*n38*
Pometia
  pinnata, 2:457*n1*, 3:163*n2*
  tomentosa, 2:457*n1*
Pommes d'Amour, 4:171
Pommom paradisi, 4:6, 4:7
Pompelmoose, 1:341*n38*, 2:136,
  2:137, 2:139–2:142, 2:163, 2:169
Pompoenen, 4:429
Pomum
  amoris, 4:169, 4:171, 4:455,
    4:456, 4:457
  Assyrium, 2:139
  aureum, 2:168
  draconum silvestre, 1:438
  hesperium, 2:168
Ponam tagera, 3:279
Pongolam, 3:593
Ponnaco Pequeno, 2:591
Poolsche mutsen, 5:443
Poolse, 5:444
Popangan, 4:569
Popeleer cussu, 5:573
Poplar, 3:47, 3:227
Poppo, 5:133
Poppya, *4:452*, 4:453–4:454
  oblonga, 4:454
  rotunda, 4:454
  utan, 4:454
  wild, 4:451
Porcupine shrub, 5:511*n4*
Porites conglomeerata, 5:445*n1*
Portulaca, 5:317
  indica, 4:215
  marina, 4:215, 5:315
  minima, 4:215
  oleracea, 2:631*n2*, 2:643*n2*,
    4:215*n1*, 4:215*n3*

quadrifida, 4:215*n5*, 5:317*n16*
rubra, 4:215
tuberosa, 5:317*n16*
Portus reticulatus, 5:441
Posocarps rumphii, 2:489*n1*
Postis iterativus, 5:122
Pot's crust, *4:530*, 4:533–4:535
Potades, 4:370
Potato, 4:171*n20*, 4:183*n3*
  sweet, *4:368*, 4:368–4:372,
    4:369*n1*, 4:461*n2*
Potherbs, 1:485, 1:485*n6*, 4:285,
  5:44, 5:53, 5:66, 5:72, 5:90,
  5:254, 5:255, 5:318
  blitum, 4:157, *4:158*, 4:159–
    4:160
  catappa, 3:101
  crackling, *4:562*, 4:563–4:564
  priest's, 4:297, *4:298*
  sour, 4:229
  wandering, 4:461, *4:462*, 4:463
Pothos
  latifolius, 3:485*n3*, 4:571*n1*
  longifolius, 4:581*n4*
  rumphii, 4:573*n1*
Poulo niry, 4:565
Poultry plant, 4:237, *4:238*, 4:239
Pousolzia zeylanica, 5:55*n1*, 5:83*n2*
Poutaletsje, 1:425
Pouteria, 3:105*n1*
Poutsjong Tree, 2:276
Pox plant, 5:115
Poxwood, 3:69
Praatjes, 3:625
Pradjittan, 4:232
Prapadeira de Cobra, 3:469
Prattlings, 3:629
  puniest, 3:635–3:636
  small forest, 3:631, *3:632*, 3:633
Prayer Beads, 3:565*n1*
Pregnant tree, 1:296*n7*
Premna
  nitida, 2:454*n3*, 3:153*n1*,
    3:389*n5*, 3:555*n3*, 3:595*n2*

obtusifolia, 3:155*n1*, 3:399*n4*, 5:242*n10*
   serratifolia, 1:496*n7*
Premna nitida, 2:350*n8*
Prickly
   broom, 3:623, 3:629*n12*
   pear, 3:331*n13*
Primrose, 4:267*n5*
Primula veris, 3:265
Princess flower, 5:181*n35*
Protium javanicum, 5:561*n1*
Provence Wood, 3:514
Prune tree, Indian, 1:388, 3:383
Prunella, 3:55, 3:55*n8*, 5:171*n23*
   alba rotunda, 5:129
   molucca
      hortensis, 5:56, 5:57–5:58, 5:129*n7*
      silvestris, 5:56, 5:59–5:61
   white, 5:99
Prunum
   cereum, 2:227, 2:228
   stellatum, 1:362
Prunus
   spinosa, 1:385*n3*
   zeylanica, 2:178
Pseudarthria viscida, 5:289*n1*
Pseuderanthemum
   curtatum, 5:63*n1*
   pulchellum, 5:99*n3*
   racemosum, 1:57, 5:103*n1*
Pseudo
   capsicum litoreaum, 3:417
   corrallium rubrum fistulosum, 5:420, 5:421–5:422
   indicum, 4:147
   sandalum
      Amboinicum, 1:454, 2:78, 2:79–2:82, 2:245
      Buronense, 2:79–2:82
Pseudochina, 3:611, 4:498, 4:503
   alba, 4:491, 4:494
      latifolia, 4:493
   amboinica, 4:491, 4:492, 4:493–4:495

molucca, 4:491, 4:492, 4:493–4:495
   nigra, 4:493, 4:494
Psidium
   cujavillus, 1:411*n1*
   guajava, 1:403*n1*, 1:407*n1*, 2:546*n11*, 2:547*n2*, 2:559*n4*, 2:593*n2*, 3:141*n3*, 3:255*n4*, 3:271*n24*, 3:395*n1*, 3:427*n6*, 3:615*n12*, 4:466*n7*, 5:109*n16*, 5:242*n9*, 5:263*n16*, 5:579*n2*, 5:579*n3*
Psilotum triquetrum, 5:173*n7*
Psophocarpus tetragonolobus, 4:381*n1*
Psychotria, 5:485*n1*
Pterocarpus, 2:79*n3*–2:5
   indicus, 1:102, 1:457*n5*, 2:55*n20*, 2:71*n4*, 2:85*n4*, 2:121*n2*, 2:135*n223*, 2:313*n1*, 2:422*n4*, 2:479*n3*, 2:496*n5*, 2:527*n2*, 2:542*n12*, 2:577*n6*, 3:11*n2*, 3:23*n2*, 3:127*n4*, 3:373*n3*, 3:499*n5*, 3:532*n7*, 3:560*n22*, 3:631*n4*, 3:669*n19*, 4:321*n25*, 4:326*n10*, 4:513*n19*, 5:251*n4*, 5:476*n5*, 5:529*n2*, 5:543*n2*, 5:561*n4*
   papuanus, 2:62*n2*, 2:319*n1*
   santalinus, 2:65*n5*, 2:71*n5*, 2:74, 3:272*n32*, 5:479*n7*
Pterospermum javanicum, 2:478*n6*
Ptychosperma appendiculatum, 1:285*n5*
Pu sang-tjan, 4:245, 4:245*n13*
Pua, 1:226
Pucki-andjing, 1:446
Pudding Pipe Tree, 2:121*n1*
Pueraria phaseoloides, 4:406*n11*
Puffballs, 5:251*n1*
Pujong, 3:223
Pula trappedera, 4:466
Pulang kean, 5:80
Pulassari
   pohon, 2:577

shrub, 3:47, 3:521, *3:522*, 3:523–3:524, *3:525*
tree, 2:577, *2:578*, 2:579
Pulassarium, 2:577, *2:578*, 2:579, 3:521, *3:522*, 3:523–3:524, *3:525*, 5:544*n1*
spurium, 3:523*n6*
Pule, 2:14, 3:320, 3:325, 3:521
Batu, 2:454
batu, 3:173
Pulesari, 2:183
Puli
marua, 3:293
puti, 3:295
Pulmonaria, 4:168
Pulo Batu, 2:374
Puloron root, 4:557, *4:558*, 4:559–4:560
Puluttan, 5:113
Pulverum Brasilianum, 4:188
Pumpkin, 4:427, 4:431, 4:432*n16*
rope, 4:555–4:556
Punanga Bug, 2:323
Punctatum, 2:59
Punica granatum, 1:391*n13*, 2:137*n1*
Punnaca, 2:325
Purslane, 3:523, 5:69, 5:315*n4*
Indian, 4:215
sand, 5:317
sea, 5:317
wild, 5:317
Pustula
arborum, *4:548*, 4:553
lactea, 4:539
Putang, 4:250
Putat, 3:450
utan, 3:101
Puti, 4:27
Putjoe pau, 4:11
Putra waly, 3:609, *3:610*, 3:611
Putsja paeru, 4:417
Putsjo pou, 1:323

Putsjoc Ana, 2:88
Putsjock, 4:63, 5:469
Pycreus odoratus, 5:4*n4*
Pyn-appel Boomen, 3:435, *3:455*
Pyrard Tree, 2:177
Pyrethrum, 5:279
Pyrum Calappa, 1:193

Q

Qua quaarer, 4:502
Qualbonen, 4:413
Qualie, 1:327
Quamoclit
coccinea, 4:466*n4*
pennata, 3:371*n2*, 4:465*n1*
Quanglang, 1:298
Quassia indica, 2:386*n27*, 3:125*n2*
Quehaan, 3:331
Quercus, 5:377
marina theophrasti, 5:381
molucca, 2:565
Quie, 3:254
Quigombo de chero, 3:244
Quince tree, 1:505*n1*
Quinua, 4:159
Quisqualis, *3:588*, 3:589–3:591
indica, 3:143*n3*, 3:589*n1*

R

Rabaco, 4:525
Rabey, 5:141
Rabna bacadja, 4:249
Radish, 4:169*n5*
Radix
amarois, 2:190, 2:191
china, 4:491, 4:495, 4:497–4:500, 4:501–4:502, 5:235, 5:593
deiparae spuria, 2:187, *2:188*, 2:189
mustelae, 5:521, *5:522*, 5:523–5:524, 5:589
puloronica, 4:556, 4:557, *4:558*, 4:559–4:560

toxicaria, 2:406, 5:299, 5:309,
   5:300301–5:305
vesicatoria, 3:359, 4:520, 4:521–
   4:522
Raiz
   da mostarda, 3:593
   de cobra, 4:560
   de mongo, 5:521, 5:523
   do boba, 4:495
Rakta Tsjandanam, 2:72
Ramak Daging, 2:261–2:262
Rambut
   papouwa, 5:165
   putri, 4:583
Rami tsjina, 4:129
Ramis, 1:283
Ramium
   majus, 4:132, 4:133–4:134
   minus, 4:128, 4:129–4:130
Rammanasses, 4:169
Ramno rubus, 3:510
Rana
   diacq, 4:519
   sanvali, 4:413
Rangas, 2:393
Rangrec, 5:207
Ranig, 1:196
Ranti, 5:119
Ranunculaceae, 5:179n2
Ranunculus, 3:269
   lingua, 5:334n7
Raphanus, 4:169n5, 4:373n5
Rapo leko, 4:326
Rappotsjidu, 1:344
Rarak Tree, 2:199–2:200
Rariorum plantarum, 3:425
Rasamalla, 2:83, 2:83n1
Rasp tree, *3:138*, 3:139
Rasphout, 3:272
Rattlepod, *4:230*, 4:231–4:232
Rauparoro, 3:301
Rauvolfia serpentina, 5:483n4,
   5:521n2, 5:523n7

Rauwolfius, 3:254
Raved seni, 5:288
Ray ray, 5:49, *5:50*, 5:51, *5:52*, 5:53
Rayat, 3:219
Razor grass, 1:190, 1:207, 2:105
Rebut, 5:301
Red Cedar, 2:569n2
Red Tree, 2:309–2:311, 2:547–
   2:548
   first, 2:543, *2:544*, *2:545*, 2:546
Reed
   arrow, 3:211, *3:212*, 3:213
   grass, 3:213
   mace, 1:350n5
   tree
      arrow, 3:189–3:190
      bullutuy, 3:189–3:190
      thorny, 3:199, *3:200–3:201*,
         3:202
      wild, 3:203, *3:204*, 3:205–
         3:207
Reed-Like Plant, *3:214*, 3:215–
   3:216
Rembuta Tree, *2:114*, *2:115–2:117*
Remirea maritima, 5:9n1
Remut kehur, 5:337
Renanthera
   coccinea, 5:191n5
   moluccana, 5:191n1, 5:197n3
Renda casian, 5:193
Rendang, 5:569
Resina
   canarina de xulabessi, 2:227
   carneola, 2:261–2:262
Rest-harrow, 5:85n1
Resta bovis, 5:85
Restiaria
   alba, 3:109, *3:110*, 3:111
   nigra, 3:113
   prima, 2:597
Retepora monilifera, 5:441n4
Reticulum
   marinum, 5:436, 5:441
   saxeum, 5:436, 5:441

Retsje, 4:186

Rex
  amarosis, 2:191–2:194, 2:375,
    3:124, 3:125–3:126, 5:520
  silvarum, 3:127

Rhamno rubus, 3:507

Rhamnus, 2:183
  frangula, 3:507*n10*
  malabaria, 3:254
  utilis, 3:508*n23*

Rheum
  rhabarbarum, 5:287*n1*
  rhaponticum, 3:535*n11*, 5:288*n9*

Rhinacanthus nasuta, 3:295n2,
  5:99*n1*, 5:282*n14*

Rhipidipathes reticulate, 5:381*n4*

Rhizophora candelaria, 2:611*n1*,
  3:434*n9*, 3:455*n2*

Rhodamnia, 3:403*n1*

Rhodophyceae, 5:339*n8*

Rhodophyta, 5:358*n13*

Rhubarb, 5:287–5:288
  Chinese, 5:287
  mountain, 5:287

Rian Rian, 3:159

Ribu ribu, 5:298

Ribua, 5:301

Rice, 4:107–4:112
  sticky, 4:113

Ricinocarpos, *3:130*, 3:131

Ricinoides, 3:287, 3:291, *3:344*,
  3:345

Ricinus, 3:117, 3:121, 3:131, 3:335,
  3:336, 3:338
  agrestis, 3:335
  albus, 5:493*n6*
  americanus, 3:339, 4:186
  communis, 1:414*n3*, 1:499*n10*,
    2:273*n4*, 2:359*n3*, 2:593*n4*,
    3:117*n6*, 3:117*n10*, 3:315*n7*,
    3:335*n2*, 3:336*n8*, 3:337*n9*,
    3:341*n1*, 3:617*n5*, 3:652*n4*,
    4:301*n3*, 4:527*n4*, 5:63*n3*
  ruber, 3:117, 3:341, *3:342*, 3:343
  syriacus, 3:341

Riddersporen, 5:189*n16*

Riedelia lanata, 5:268*n9*

Rihut, 5:273

Rima tehu, 1:420

Rinat, 1:473

Ringan ringan, *3:368*, 3:393

Ringworm tree, *5:530*, 5:531–*5:532*

Rite, 2:374, 2:378, 2:406

Ritsjes, 4:183, *4:184*, 4:185–4:189

Riud, 4:27, 4:29
  lacki lacki, 4:29, 4:30
  mera, 4:29
  puti, 4:29

Rizo, 4:110

Robinsoniodendron ambiguum,
  3:388*n10*

Robong, 3:188

Rock flower, 5:170

Roda de China, 3:224

Roffu, *3:398*, 3:399

Rolled-Up Flower, *3:324*, 3:325–
  3:326

Romanca Tree, 2:137

Rombia, 1:299

Romiho, 1:299

Rompot
  andjing, 5:19
  assap, 5:19
  bantal, 5:165
  bebeck, 5:22
  bulu
    babi, 5:31, i5:32
    matta, 5:10, 5:34
  carbou, *5:16*, 5:17–5:18
  cras, *5:16*, 5:17–5:18
  cuda, *5:25*, *5:26*
  dingin, 5:171
  djintam, 5:31, i5:32
  katjano, 5:105
  majara, 5:19–5:20
  manni manni, *5:40*, 5:41
  matta ican, *5:13*, 5:15
  negri, 5:49, *5:50*, 5:51, *5:52*, 5:53
  ombong, 5:21–5:22, 5:21*n5*
  siree, *5:26*, 5:29–5:30

Ronde weeg-boonen, 3:568

Roode, 4:401
  nieboom, 1:242

Root Tree, 2:601

Rope
  bilious, 5:540
  fat, 5:544
  muraene, 5:540
  plant, *3:496*, 3:497
    quadrangular, *3:472*, 3:473
    shrub
      burning, *3:488*, 3:489–3:490
      gnemon-shaped, *3:484*, 3:485
      osier-like, *3:470*, 3:471
      rolled-up, *3:584*, 3:585–3:586
    stream, 5:371

Roppa rappo java, 1:387

Ropy Clompan shrub, 3:587

Roraco, 3:137

Rosa
  canina, 2:173, 2:173*n5*
  rubiginosa, 3:269*n2*, 4:491*n4*

Rosa Batavico-Indica, 3:221

Rosary herb, *5:40*, 5:41

Rose
  apple, 1:369*n1*
  apple tree, 2:207, 2:208, 2:209–
    2:210, 2:211
  mallow, 3:217*n1*
  sinica, 3:224

Rosemary, 4:141, 4:144, 4:145,
    4:257, 5:281
  Ambonese, *5:46*, 5:47–5:48
  wood, 2:79

Rosmarinus officinalis, 1:411*n2*,
    5:48*n6*

Rottang, 3:637, *3:638*, 3:639–3:642,
    5:371
  alea, 3:645, *3:646*, 3:647–3:648
  bezaar, 3:639
  bulu, 3:670, 5:540
  calappa, 3:643
  cawa
  djernang, 3:664, 3:666, 3:667,
    3:668

itam, 3:639, 3:643, *3:644*, 3:661,
    3:664, 3:667, 3:669–3:670
java, 3:639, 3:655
laut, 5:369*n1*, 5:371
oëtan, 3:673–3:676
puti, 3:639, 3:664
sour, 3:669–3:670
toëni, 3:667
tsjavoni, 3:639, *3:654*, 3:655–
    3:656, 3:657, *3:658*, 3:659–
    3:670
tubu, 3:643
tuni, 3:639, 3:649, *3:650*, 3:651–
    3:653, *3:654*, 3:655–3:656,
    3:669
utan, 3:661, 3:671, *3:672*
wanitette, 3:656
zalack, 3:661, *3:662*, 3:663

Rottinga, 5:13

Roughsack, 1:344

Round flower, *4:246*, 4:249–4:250

Routsjaga, 5:71
  goffo, 5:71
  gumi, 5:71

Roweyl, 2:81, 2:245, 2:246

Royal
  boll, 5:233–5:235, 5:246
  tree, 2:387, *2:388*, 2:389–2:390
  tuber, 5:233*n7*
  upas, 2:401

Rricinus communis, 4:584*n14*

Ruba baccifera, 3:171*n5*

Rubus
  fraxinifolius celebicus, 3:621*n1*
  integrifolius, 3:8
  moluccanus, 3:8*n46*, 3:620,
    3:621–3:622, 3:621*n3*
  parvifolius, 3:621*n1*

Rucam femina, 5:533, *5:534*, 5:535

Ruccu ruccu, 4:209

Rucu, 5:5, 5:25, 5:26

Rudens
  silvaticus, 3:606, 3:607–3:608

Ruellia repens, 3:295n7, 5:101*n1*

Ruha, 3:407

Ruma
  codoc, 5:257
  kricki, 3:427
  sumot, 5:229, _5:230_, 5:231
  yar waccar, 5:339
Rumbadu, 3:5
Rumex, 3:245n4
  acetosa, 4:527n3
  patientia, 4:229In6
Rumphelia antipathes, 4:154n8,
  5:361n3, 5:397n10
Rumut, 5:157
Run-boom, 2:604
Runun Laumuri, 2:587
Ruscus, 5:311n1
  aculeatus, 5:311n1
  silvestris, 5:311
Russok, 3:270
Russu layn, 3:273
Russukey Tree, 2:293
Ruta chalepensis, 5:349n2
Rutton, 1:372
Ruttu
  layn, 1:375
  oär, 1:375
  pya, 2:559
  ruttu, 5:163, _5:164_, 5:165
    caju, _5:172_, 5:173–_5:174_
    laut, 5:381
    papuwa, 5:165
    parampaean, 5:165
  tsjapeo, 1:377
Ruun, 1:372
Ryf-boom, 3:406
Ryst, 4:110
Ryst-korls-pinang, 1:244

S

St. Maria's Root, 2:183, _2:184_,
  2:185–2:186
  bastard, 2:187, _2:188_, 2:189
St. Thomas bean, 1:31, 3:475n1
Sa pohon pissiang, 4:6
Sabatimbia, 3:411

Sabe, 1:495
Sabika, 2:559
Sabina, 2:569
  indica, 2:570
Sabrang, 4:377
Sacar
  bebec, 3:293
  belek, 3:294
Sacca, 3:34
Saccharum
  edule, 4:93n1, 4:99n1
  officinarum, 1:278n26, 2:133n7,
    3:435n19
  spontaneum, 3:211n2
Saccus
  arboreus
    major, 1:344
    minor, 1:347, 1:348, 1:350n1
Sack sack, 4:470
Sad
  glad plant, 5:283, _5:284_
  plant, _5:78_, 5:79–5:81
  tree, 3:559n11
Safflower, 1:337n14, 4:123n5
Saffran de Terra, 4:60
Saffranici flores, 4:136
Saffron
  earth, 4:60
  flower, 4:136n10
  meadow, 5:213, 5:309n6
Sagarum maleficia, 5:60
Sage, 3:255
  East-Indian wild, 5:108n12
Saggina, 4:105
Sago palm, 4:303n8
Sagu, 1:206, 1:213, 1:272, 1:292,
  1:293–1:296, _1:297_, 1:298–1:300
  duri-rottang, 1:296
  manta, 5:171n21, 5:171n22
  parampuan, 1:298
  shrub, 3:637
  smooth, 1:298n22
  tree, 3:445
  tree sajor, 4:454
Sagua viro, 4:51

Saguaster, 1:242, 1:247, 1:281
    major, 1:283
    minor, 1:287

Sagueer tree, 1:273, *1:274*, 1:275–
    1:280, 3:258, 3:258*n13*, 3:409*n8*
    wild, 1:281, *1:282*, 1:283–1:284,
        1:285, *1:286*, 1:287

Sagueiro, 1:275

Saguer plaister, 5:259

Saguerus, 1:273
    femina, 1:273
    silvestris, 1:283

Sagumanta, 2:83

Sagus, 1:299
    filaris, 3:453, 3:647
    genuina, 1:296
    laevis, 1:298
    longi-spina, 1:298
    silvestris, 1:298

Saguweer tree, 1:212*n19*

Saha saha, 3:349, 5:19–5:20

Sahit metten, 3:603

Sahu naccur, 4:291

Saia, 1:460

Saihe maruani, 4:494

Saja
    baki, 5:181, 5:181*n35*, 5:195
    balitsiajaor, 3:223
    cananga, 2:296
    cotcle, 3:563
    mamj, 3:359
    manorae, 3:558
    ngalli ngalli, 3:223
    ngawa, 5:181, 5:195

Sajo hollanda, 4:131

Sajob bandan, 4:297

Sajor, *1:494*, 1:495–1:496, 1:497,
    1:498, 1:499
    Ambonese, 4:511
    assam, 4:229
    babi, 4:297, 5:65
    bajang, 4:159, 4:160, 4:161,
        4:191, 5:44, 5:53, 5:69, 5:83*n4*,
        5:95, 5:99
    bangala, 4:129
    beach, 4:545

bengala, 5:69
bintor, 5:331
boom, 1:485
bulu, 4:487, 5:67
    ajer, 5:44, 5:45
    besaar, 5:44
    kitsjil, *5:40*, 5:43–5:45
caju, 5:241
calappa, 1:311, *1:312*, 1:313–
    1:314, 1:314, *1:315*, *1:316*,
    1:317, 1:317, *1:318*, 1:319–
    1:320, 1:320, 5:123*n13*, 5:298
    lacki lacki, *1:318*
    utan, 1:313
cambing, 3:153, 5:103
canda gondic, 5:44
carang, 5:339
catappan, 3:101
codock, 5:331
gebe, 3:239
haram tsjada, 5:44, 5:67
Javanese, 5:115
kuyt, 4:99
laut, 5:170
lomba, 5:259
lompan, 5:337, 5:345
maccou, 1:499, 1:511, 4:545
    darat, 4:544
mamma, 4:234
manis, 5:298
muca manis, 4:261
nuno, 3:16
pacu, 4:471, 5:127, *5:128*, 5:129–
    5:130, 5:130, 5:131
    babulu, 5:130
pangayu, 5:295
panu, 5:93
pepe, *3:596*, 3:597–3:598
poppya, 4:454, 4:544
preto, 4:237
pti pulo, 1:497
puti, 5:193, 5:241
radja, 1:313, 5:153, 5:155
songa, 3:350, *4:468*, 4:469–4:472
tana, 5:243
telloorican, 4:99
truba, 3:435, 3:435*n19*
trubu, 4:99
udang, 5:69
wassa, 3:41

Saketa, 2:271

Sakka, 3:35, 3:37, 3:38
  Javana, 3:37

Sala, 5:89

Sala Sala, 2:601

Salacca edulis, 3:661*n1*

Salackal Tree, 2:96

Salamander tree, 3:147

Salamgaro, 3:247

Salattan, 3:176

Salattene, 3:176

Salawat, 4:207

Salay, 3:402, 3:421

Salayer, 3:401

Salecong, 5:147

Salee, 4:101, 4:102
  utan, *5:40*, 5:41

Saley Lau Ila, 2:555

Salgeyra
  Ferma, 2:604
  Major, 2:604
  Mamato, 2:604

Salicornia, 5:315*n7*, 5:317,
  5:317*n13*, 5:411

Saligi, 1:247

Salimuri, 2:333, 2:340, 2:344,
  2:346, 2:366

Salisou, 3:468

Salissa, 1:466

Saliu Tree, 2:137

Salla bulo, 4:109

Salle utan, 4:101

Salo
  Baboda, 2:266
  Coclu, 2:242
  Colano, 2:268
  Garo, 2:257

Salpicado, 4:11*n25*

Salsa de mar, 5:318

Salsa do praja, 4:486

Salsify, 4:367*n15*

Salsula marina, 5:318

Salt Leaf Tree, 2:627

Salt Wood, Small, 2:515

Salua, 3:373

Salungeri, 2:344

Salvia
  indica, 4:168, *5:106*, 5:107–5:109
    arborescens, 5:109
    odore, 5:108
    officinalis, 2:173*n3*
    sclarea, 4:251*n2*

Samacka, 1:460

Samadera indica, 5:245*n10*

Samae, 3:118

Samama, 1:277, *2:466*, 2:467–2:468

Samanca, 4:432

Samar
  Ayer, 2:446
  Batu, 2:446

Samarin Tree, 2:195

Samay, 4:73

Sambali, 3:257

Sambaruco, 4:470

Sambeng, 4:297

Sambi, 1:429

Sambodja, 3:325

Sambong, 4:581
  tulan, 4:507, 4:561

Sambu Tree, 2:209

Sambucus nigra, 3:220*n30*

Sambuta, 2:363

Sami, 1:247

Sammana, 3:593

Sampacca, 2:301, *2:302*, 2:303–
  2:305, 3:559
  alba, 2:306, 2:307
  montana, 2:310–2:311
  silvestris, 2:306, 2:307

Samstravari, 1:378

Sanbosu, 5:359, 5:359*n18*

Sanco idong, 5:49, *5:50*, 5:51, *5:52*,
  5:53

Sandalum, 2:64, 2:65–2:82
  Aethiopicum, 2:75
  album, 2:67
  citrinum, 2:67
  pallidum, 2:67
  radicis, 2:69
  rubrum, 2:71–2:77
Sandalwood, 1:42, 1:432*n14*, 2:63,
  2:64, 2:65–2:82, 2:91, 5:416,
  5:471, 5:479*n6*
  bastard Ambonese, 2:78, 2:79–
    2:82
  blood, 2:72
  common, 2:71
  Ethiopian, 2:73
  Malayan, 2:71, 2:74
  Portugeuse, 2:71
  Redware, 2:71–2:77
Sandel-wood, 3:271, 3:272,
  3:320*n6*, 3:508
Sandori tree, 1:451, 1:452, 1:453–
  1:454
Sandoricum, 1:451
  indicum, 5:565*n2*
  koetjape, 1:451*n1*
Sangius, 2:207, 2:208, 2:209–2:210,
  2:211
  draconis, 2:74, 2:318, 2:381,
    2:382, 2:383–2:386, 3:314,
    3:499, 3:500, 3:663, 3:664,
    3:666, 3:668–3:669
  foemina, 2:210
  mas, 2:209
Sanho suy, 5:439
Sanicula, 4:525
Santalum album, 1:363*n22*,
  1:432*n14*, 2:65*n1*, 2:71*n3*,
  3:272*n32*, 3:501*n9*, 4:35*n17*,
  4:62*n29*, 5:471*n14*
Santanhöa, 3:359
Santenu, 2:374
Santjone, 3:340
Santoor, 1:451
Sapal lauhali, 5:481
Sapela, 5:481
Sapindaceae, 5:491*n1*

Sapindus rarak, 2:199*n1*
Sapiri Tree, 2:273
Sapotaceae, 2:367*n13*
Sappa lalawa, 3:11
Sappal Tree, 2:597–2:598
Sappan wood, 3:57, 3:192, 3:266,
  3:267, *3:268*, 3:269–3:272, 3:273,
  3:499, 3:499*n1*, 3:508, 3:623
Sappar, 3:109
Sapussi, 4:556
Sara brassa, 1:395
Sarang burung, 5:343*n16*
Sarangan, 2:567
Sarcanda Tree, 2:70
Sarcanthinae, 5:206*n29*
Sargassum, *5:348*, 5:349–5:351
  aquifolium, 5:346, 5:346*n10*
  flavifolium, 5:346*n23*
  granuliferum, 5:346*n13*
  natans, 5:349*n1*
  polycystum, 5:345*n1*
Sari, 4:136, 4:137, 4:139
  daun, 4:326
  devil's, 3:546*n10*
  largest, *3:544*, 3:545–3:546
  luning, 2:379
  magician's, 3:546*n11*
  swangi, 3:546
  tallan, *3:544*, 3:545–3:546
  wild, 3:546*n8*
Saribu, 1:247, *1:248*, 1:249–1:250,
  1:251, 1:279
Saribus, 1:247
  pumila, 1:250
Sarile halat, 3:311
Sarong
  burong, 5:343
  tangan laut, 5:455
Sarsa parilla, 4:486, 4:495, 4:497–
  4:500
Sasacke Sehy, 3:139
Sasawi
  ambon, 4:235
  tsjina, 4:235–4:236

Sasicka, 4:6

Sassafras, 2:96
   albidum, 2:96n7

Sassala, 4:146

Sassea Tree, 2:281

Sassise Niwel, 2:575

Sasulen, 5:122

Sasupite, 3:571

Sasuru, 1:454, 2:79, 2:393, 3:333
   Utan, 2:370, 2:393

Satore, 3:477

Sattul tree, 1:451, *1:452*, 1:453–
   1:454

Saturn's beard, 5:167

Satyrium, 5:179, 5:201n1

Sauce Leaf, 3:141–3:142

Saun idop, 4:268

Saururus, 3:543

Sawaccal, 5:243

Sawan Tree, 2:276

Saxifrage
   English, 5:140n13

Saxifragia Anglorum, 5:140

Saxum
   aquosum, 5:449
   calcariium marinum, 5:445–5:448

Saya tsjina, 3:266

Sayor
   ayam, 4:237
   Bayang, 2:406
   boom, 1:495
   cancong, 4:461, *4:462*, 4:463
   Catappan, 2:207
   kambing, 1:57
   maccou, 4:563
   puti, 1:495
   tawar, 4:569

Sayor bulu, 5:43–5:45

Scabiosa
   iavana, 4:262n15
   indica, 4:262

Scaevola frutescens, 3:349n2,
   3:381n2

Schadida calli, 3:333

Schaid, 3:629

Schakeri schora, 4:429

Schalêmba-Paina, 1:225

Schandulae foliaceae, 1:299

Schanga cuspi, 3:564

Schefflera, 1:420n21, 5:477n1

Schembra valli, 4:519

Schemparite, 3:220

Schera
   punca, 4:146
   schunda, 4:176

Scherab, 1:194

Scheru Cadelari, 5:53

Schir, 1:194, 1:257

Schismatoglottis calyptrata, 4:297n1

Schit thuni, 4:494

Schit-elu, 4:122

Schizaea dichotoma, 5:174n9

Schizomeria serrata, 2:546n6,
   3:119n1, 3:399n2

Schizostachyum
   blumel, 3:189n1
   brachycladum, 3:173n4, 3:187n1
   lima, 3:670n24

Schleichera oleosa, 1:432n14,
   3:559n14, 4:94n2

Schoenantum, 1:277
   Amboinicum, 4:82, 4:85–4:88
   fatuum, 1:277
   indicum sterile, 4:86

School Wood Tree, 2:372, 2:373–
   2:375

School-hout, 2:374

Schorzakken, 1:344

Schundapana, 1:283

Schwarzwurzel plant, 1:133

Schylla, 5:307

Schyrara, 4:55

Schytnootjes, 3:345

Scindapsus marantifolius, 4:579n9

Scirpiodendron ghaeria, 3:463*n1,*
5:38*n14*

Scizomeria serrata, 5:499*n6*

Scleria
bancana, 5:37*n1,* 5:223*n17*
lithosperma, 5:37*n2*
multifoliata, 5:37*n1*
scrobiculata, 5:37*n1*
sumatrensis, 5:37*n1,* 5:37*n4*

Scleroderma aurantium, 5:237*n1*

Scolopendria
major, 5:154, 5:157–5:158
minor, 5:154, 5:157–5:158
tectorum, 5:154, 5:157–5:158

Scoparia, 5:86

Scorzonera hispanica, 4:367*n15*

Scouring Leaf, 3:405

Screw beans, 5:527

Scrotum cussi, 5:573

Scutellaria, 2:80
crispa, 3:309, 3:310, 3:311
primam, 3:301, 3:302, 3:303
secunda, 2:593
angustifolia, 3:305, 3:306–
3:307, 3:308
latifolia, 3:305, 3:306–3:307,
3:308
tertia, 3:309, 3:310, 3:311

Sea
accabaar, 5:414
arrow, 2:406, 5:422
basta, 5:450, 5:451
beans, 3:474, 3:475–3:478
beard, 5:457
broomstick, 5:422*n13*
calappa, 5:384
Cape Flower, 2:291*n2*
casuaris tree, 5:379
cau, 3:90
clove, 3:368*n5*
cloves, 5:436, 5:439–5:440
coconut, 2:75, 5:357*n10*
cypress, 5:377, 5:381
fan, gray, 5:407–5:408
fans, 5:374, 5:375–5:376, 5:378
fir, 5:377*n6*
fucus, 5:340

funnel, 5:452, 5:453
glove, 5:452, 5:455
grass, 5:337*n3*
gypsies, 5:346*n19*
hay, 5:378, 5:379
heather, 5:381
holly, 5:311*n2*
hornet's nest, 5:459
jelly, 5:459
kalappa, 5:383, 5:385, 5:386
lentil, 5:349*n6*
net, 5:436, 5:441
oak, 5:377*n5*
parasol, 2:639
parsley, 2:643, 5:315*n2,* 5:315*n5*
pen, 5:422*n12*
pepper, 3:542, 3:543
pinks, 5:436, 5:437–5:438
pomegranates, 2:582
purslane, 5:317
root, 5:357*n3*
rottan, 5:371*n6,* 5:409–5:410
rush, 5:370
shade tree, 2:638, 2:639
shrub, 5:405
tree, 5:357–5:360
Indian, 3:383
tremella, 5:331, 5:340
urchin, 5:422*n11,* 5:445*n5*
wasps, 5:459

Seappu, 4:361

Seaweed, 5:349
coran, 5:339*n7*
edible, 5:339*n2*

Sebesten, 3:49

Secul, 1:451

Securidaca, 4:141*n4*

Sedge, 5:33*n3,* 5:38*n8,* 5:223*n17*

Sedum, 4:227*n3*
arborescens, 5:199

Segol tabana, 4:309

Sehit, 3:625
ela, 3:629
laut, 3:631, 3:632, 3:633

Seho, 1:273

Sehy, 4:55

Sekala, 5:273

Sel, 1:495

Selaginella
　durvillei, 5:163*n2,* 5:381*n3*
　plana, 2:11*n5,* 5:163*n1*

Selago, 5:174

Semecarpus
　anacardium, 1:471*n1*
　cassuvium, 2:391*n10,* 3:297*n2*
　forstenii, 1:474*n9,* 3:213*n6*

Sempervivum, 3:293, 3:541, 4:227,
　4:227*n1,* 4:547*n2,* 5:197
　majus indicum, 3:541*n5,* 4:221,
　　4:222, 4:223–4:225
　marinum, 4:223
　tectorum, 4:144*n14,* 5:195*n2*

Semsem, 4:119

Sene rubus, 3:270

Senecio, 5:64
　amboinicus, 5:129
　　5:64,5:67
　sonchifolius, 4:261*n1*

Senna, 4:231*n6*

Senne tree, bastard, 3:267

Sensitive plant, *4:264,* 4:267–4:272,
　5:443*n3*

Sente, 4:300

Sentigi, 2:631

Seram, 1:39
　laut, 1:38

Sergeant's balls tree, 1:334

Sergiants-klooten, 5:262

Seriatopora, 5:430*n10*

Sericum marinuum, 5:337–5:338

Sero wakki, 1:352

Serpenticida, 5:589–5:591

Serratula amara, 4:523, 4:529,
　4:530, 4:531

Seru, 3:315

Seru Seru, 3:188

Serune, 4:200
　ajer, 4:201, 4:470
　sky-blue, 4:251
　tschina, 4:470
　water, *4:468,* 4:469–4:472

Seruneum aquatile, 4:468, 4:469–
　4:472

Seruwary, 3:473

Sesame plant, 4:121*n13*

Sesamum
　album, 4:122
　Indicum, 4:117, *4:118,* 4:119–
　　4:122
　nigrum, 4:122
　orientale, 4:117*n1*

Sesbania
　cannabina, 3:281*n5*
　grandiflora, 1:441*n7,* 1:489*n1,*
　　1:492*n13,* 3:135*n4,* 3:281*n3,*
　　3:542*n16,* 3:608*n9,* 4:15*n16,*
　　4:237*n3,* 4:511*n6,* 5:543*n4*
　sesban, 3:267*n22,* 3:281*n1*

Sesel, 2:521, 2:522

Sesoot, 1:277, 2:201, 2:202, 2:203–
　2:204

Sesuvium portulacastrum, 2:643*n3,*
　4:215*n9,* 5:315*n3*

Setaria
　italica, 4:115*n2*
　parviflora, 5:34*n14*

Sewet tree, 5:476

Sgarosuti, 5:414

Shaddock, 1:341*n38*

Shield tree, 3:87, *3:88, 3:89,* 3:90

Shoe Flower, 3:219

Shore Reed, East-Indian, *3:208,*
　3:209

Shorea
　leprosula, 2:261*n5*
　selanica, 1:424*n8,* 2:88*n42,*
　　2:233*n4,* 2:253*n1,* 2:256*n4,*
　　3:667*n17,* 5:476*n2*

Shrub of marbles, 3:623, *3:624,*
　3:625–3:626

Siaelaut, 3:135

Siaffu, 4:354

Siaho, 4:354

Siamese flower, *5:570, 5:571*

Siawaro, 4:39

Sibac langaa, 4:454

Sibane, 5:211

Sibut Tree, 2:273

Sicappa, 4:361

Sicchius, 2:478

Sicki Tree, 2:475, 2:476, 2:477, 2:478

Sico Hajate, 2:281

Sida
  acuta, 5:85*n4*
  alnifolia, 3:230*n8*, 5:85*n2*
  cordifolia, 3:230*n10*
  rhombifolia, 5:85*n2*
  sonneratiana, 1:31, 3:231*n16*
  triloba, 1:31, 3:231*n16*

Sidagori, 5:513

Sideritis, 5:75
  quarta, 5:73

Sideroxylon, 2:475*n4*, 2:519*n1*
  microcarpum, 2:424*n10*, 2:549*n1*

Sidium quajava, 1:377*n2*

Siëota, 3:100

Siger, 1:265

Sigillum salomonis, 5:297

Sija mahuri, 3:414

Sikat, 3:87

Sikki tree, 3:159

Sikkoa, 4:432

Silagurium, 5:82, 5:84, 5:85–5:87

Siliqua Aegyptia, 2:123

Siliquastrum, 4:186

Silk cotton tree, 1:335*n6*, 1:501*n1*

Silkweed, 5:101*n2*

Sillicypria, 3:336

Silva de praja, 4:272

Simbar, 5:150, 5:153, 5:157
  betul, 5:151
  layangan, 5:150
  majangan, 5:155

Simpliciter serrata, 3:507

Sinupi
  Amonicum, 4:235
  juncea, 4:235*n2*
  sinense, 4:235–4:236

Sinapister, 3:592, 3:593

Sinapistrum, 4:234

Sindoc Tree, 2:96, 2:98, 2:99

Sindora galedupa, 2:85*n1*, 2:304*n17*, 3:501*n20*

Sindric maul, 4:193

Sineese
  boontjes, 4:383
  camillen, 4:200
  mater, 4:200
  peren, 1:395

Singhasari, 1:35, 1:36

Singi Tree, 2:542

Singomaas, 4:379

Sinlo som, 5:548

Sion, 4:525

Sipa sipa, 3:451

Sirea, 1:277, 1:495

Siree, 4:86, 5:29
  utan, 5:32, 5:35

Siri, 4:189
  andjing, 3:542, 4:331
  bertemmu, 4:320
  boa, 3:321
  boppal, 3:45
  boppar, 3:39, 3:43, 3:45, 3:46, 3:118, 3:118*n17*, 5:105
  candhati, 4:329
  caya, 1:404
  culilawan, 4:325
  daun, 3:78, 3:541, *4:314*, 4:319–4:323, 4:320, 4:323, 4:335, 4:353
  folia, 2:520, 2:521, 2:522, 2:523
  folium, *4:314*, 4:319–4:323
    silvestre primum, *4:328*, 4:329
  java, 3:549, 4:319, 4:333
  leaf, 4:320
  pangel pangel, 4:316
  payt, 3:542
  pinang, 4:84, 5:555
  seytan, 4:332

tana, 4:331
third tree-like, 3:547, *3:548*,
  3:549–3:550
tsjatsjin, 4:331
tsjencke, 4:320
utan, 3:531, 3:547, *3:548*, 3:549–
  3:550, *4:328*, 4:329
  babau, 3:547
  dingin, 4:333
  wild, *3:540*, 3:541–3:543
Siri-Leaf Tree, 2:520, 2:521–2:522,
  2:523
Siriboa, 1:295, *4:314*, 4:319–4:323,
  4:324, 4:325–4:327
  ayer, 4:325
  cambing, 4:325
  fragrans, 4:325
  puti, 4:325
  silvestris, 4:329
Siricaya Tree, 2:169
Siricaynona, 1:395
Sirikaya, 1:399
Sirioides alter, 3:554
Sirium
  arborescens tertium, 3:547, *3:548*,
    3:549–3:550
  decumanum, 3:544, *3:545*–3:546
  frigidum, 4:333, *4:334*
  silvestre, 2:521, 4:329*n2*
  terrestre, *4:330*, 4:331–4:332
Siroides, 3:552, 3:553–3:554
Sisi seytan, 3:546
Sissin, 3:403
Sisymbrium officinale, 4:233*n3*
Siu Hunaë Tree, 2:281
Sium
  latifolium, 4:523*n7*
  sisarum, 3:313*n3*
Siwe
  anahan, 4:51
  lale, 4:55
  tanahan, 4:51
  yla, 4:51
Sjampr ingat, 3:425
Sky tree, *3:148*, 3:149–3:150
Slackhout, 2:351

Slime apple, *1:508*, 1:509
Slimy Wood tree, 3:143, *3:144*
Sloe, 1:385*n3*
  East Indian, 5:569
Slymappels, 1:506
Small Palm, 1:201*n4*
Smaltkruyt, 3:564
Smilax, 4:335*n2*, 4:491, 5:592,
  5:593–5:594
  aspera, 4:491, 4:495, 4:497, 4:499
  australis, 4:491*n2*
  china, 2:389*n14*, 3:660*n15*,
    4:262*n18*, 4:497*n1*, 4:501*n1*,
    5:235*n30*, 5:593*n1*
  javensis, 3:611*n14*, 4:491*n2*
  leucophylla, 4:493*n9*, 4:493*n11*
  pseudo-china, 5:235*n27*
  regelii, 4:497*n1*
Snail-wood, 2:351*n18*
Snake
  herb, 5:63
  plant, 5:105*n10*
  slime plant, 5:60
  tongue, 4:367*n14*
  tongue plant, 5:147*n2*, 5:292,
    5:295, 5:297–5:298
  ubi, 4:343
Snakewood, 2:65, 2:179, *2:180*,
  2:181–2:182, 3:554, 4:560
  Ceylonese, 3:300
Snygras, 5:33*n2*, 5:33*n3*, 5:223*n17*
  Inidaans, *5:32*, 5:33–5:34
So-ulo, 4:73
Soa
  abbal, 3:487
  sapyro laut, 5:422
  wale, 3:485
Soam sueng, 4:470
Soap Bark, 3:413, *3:414*, 3:415
Soap-Balls Tree, 2:199–2:200
Sobe leypossi, 3:674
Soccan capas, 1:352
Soccum, 2:446, 3:175, 3:176
  batu, 1:352, 1:357, 3:41
  bidji, 1:356, 1:357

capas, 1:356
lanosus, 3:118
maas, 1:352
pissang, 1:352
radja, 1:352
seeded, *1:354*, 1:355–1:356
telloor, 1:352
utan, 1:356, 1:357
wild, 1:357, *1:358*, 1:359, 3:109

Soccus
bidji, 1:350*n3*, 1:351, 1:355
capas, *1:350*, 1:350*n2*, 1:351,
1:352, 1:355
granosus, 1:350*n3*, 1:356
lanosus, 1:350*n2*, 1:352
silvestris, 1:350*n4*, 1:357
utan, 1:355

Sochus amboinicus, *4:260*, 4:261–
4:262

Socur, 4:73

Soffera, 3:267

Soffo
manni manni, 3:387
mutiara, 5:67

Solamu lohon, 3:275

Solanum, 4:191
album, 4:175*n2*
*bacciferum
esculentum*, 2:207, 5:259
furiosum, 3:528
indicum esculentum, 4:169, 4:170,
4:171
maniacum, 4:177, 4:179
melongena, 2:207*n3*, 3:607*n3*,
4:19*n6*, 4:169*n3*, 4:169*n7*,
4:169*n12*, 5:259*n5*
nigrum, 5:119*n1*
odoriferum, 4:192
pomiferum, 4:171, 4:455
pressum, 4:175*n3*, 4:176
spinosum, 4:173
trongum, 4:173*n1*
verbascifolium, 5:72*n12*, 5:111*n1*
vesicarium, 5:117

Soldanella, 4:415, 5:317*n13*
marina, 4:413, 4:486
indica, *4:484*, 4:485–4:486

Sollio coupan, 5:531

Solomon's seal, 5:297*n4*

Solulo, 2:495

Solulus, *3:136*, 3:137
arbor, *3:136*, 3:137
funicularis, 3:542, 3:608
pohon, 3:137

Som, 5:545, *5:546*, 5:547–5:553

Sombang, 3:5

Somboc, 3:314

Sombong, 5:107, 5:108, 5:111
bankey, 5:107, 5:108

Somliuw, 5:548

Sonchum laevum, 4:262

Sonchus
amboinicus, 1:266, 5:129
oleraceus, 4:261*n6*
volubilis, 4:263, *4:264*, 4:265
Javanus, 4:265

Soncorus, *4:70*, 4:73–4:74

Soncory plant, *4:70*, 4:73–4:74

Sondal malam, *4:240*, 4:241–4:242

Songi Tree, 2:209, 2:210

Sonneratia, 2:547*n6*, 2:582*n9*
alba, 1:389*n3*, 4:513*n18*,
5:140*n11*
caseolaris, 2:519*n2*, 2:617*n5*,
3:118*n14*, 4:551*n4*, 5:150*n9*,
5:193*n3*, 5:313*n9*

Soo, 1:477

Soorzak, *1:342*, 1:344, 1:345

Sophocarpus tetragonolobus, 5:531*n4*

Sophora
japonica, 2:379*n5*
tomentosa, 2:192*n13*, 2:386*n25*,
3:125*n4*, 3:271*n29*, 3:273*n1*,
3:619*n8*, 5:51*n11*, 5:281*n6*,
5:479*n3*, 5:569*n3*

Sorgelin, 4:119

Sorghum
amboinicus, 5:35*n1*
bicolor, 4:103*n3*

Sorghum seed, 4:103, *4:104*, 4:105

Sorrel, 3:517, 4:489*n2*, 4:527*n3*
  Indian, 3:247
  Spanish, 4:229
  wild, 3:245
  wood, 4:269*n16*

Sosillo, 3:387

Sosoru, 5:89
  bobuto, 5:91

Sosuhu, 3:329

Sotectsou, 1:320

Soteetsjou, 1:319, 1:320

Soulamea amara, 2:191*n1*,
  2:375*n14*, 2:386*n26*, 3:125*n1*,
  3:325*n2*, 3:327*n5*, 3:535*n9*,
  4:446*n12*, 5:245*n9*

Soulamou, 3:125, *3:540*, 3:541–
  3:543, 3:554, 5:519, 5:519*n2*
  ambon, 3:535
  pohon, 3:325
  tali, 3:542

Soulamu, 2:192

Sour leaf, 4:527*n8*
  large, 2:509, 2:510, 2:511
  small, 2:512, 2:513–2:514

Soursack, 1:344

Soursop, 1:343*n1*, 1:397*n12*

Soutenelle, 5:317, 5:411

Souterig, 5:317

Southernwood, 5:429*n6*

Sow plant, *5:64*, 5:65–5:66

Spaansch speck, 4:421

Spandia dulcis, 3:254*n37*

Spanish Cedar, 2:569*n2*

Spanish galingale, 5:11*n5*

Spartium, 5:33, 5:33*n6*

Spathoglottis plicata, 5:213*n3*,
  5:223*n7*

Spattenboom, 2:400

Spatwortel, 4:273*n2*, 5:301

Spearmint, 5:291

Spearwort, 5:334*n7*

Speedwell, 4:529*n2*, 5:291*n2*

Spek mangas, 1:325

Speldewerk, 5:446

Spelten-gras, 5:23, 5:24

Spina
  alba, 5:313
  spinarum, 5:533, 5:534, 5:535
  vaccarum, 3:503, 3:504

Spinach, 4:461, 4:462, 4:463

Spinifex littoreus, 5:10*n9*

Spiralis
  hirsuta, 5:275, 5:276, 5:277
    leavis, 5:277
  leavis, 5:275, 5:276, 5:277

Spleenwort, 5:140*n14*, 5:157*n2*
  maidenhair, 5:145*n2*

Splinter tree, *3:70*

Splits-Hoorens, 2:617

Spondias
  amara, 5:465*n7*
  cythera, 2:385*n23*, 2:528*n7*
  cytherea, 4:62*n25*, 4:68*n11*,
    4:140*n46*
  dulcis, 1:379*n7*, 1:439*n1*, 2:369*n2*
  pinnata, 1:443*n1*

Spongia
  elegans, 5:453
  infundibili forma, 5:452, 5:453

Spotted plant, 5:92, 5:93–5:94

Sprig of stone, 4:528*n10*

Spurge, 5:465*n6*
  sun, 5:105*n4*

Square rope plant, 3:611

Squilla, 5:307*n2*, 5:308*n7*

Stallkraut, 5:85*n1*

Standelwort, 5:213*n2*

Standergrass, 5:213*n2*

Staphyledendron indicum, 1:429

Steene moernagels, 5:439

Stellar prune, 1:362*n6*

Stellaria, 5:43*n1*

Stemona
  asperula, 4:366*n6*
  moluccana, 3:350*n12*, 4:365*n1*,
    4:366*n6*, 5:298*n11*

tuberosa, 4:365*n1*, 4:494*n14*,
    4:506*n9*, 5:298*n9*
Stemonurus, 3:69*n1*
Stenochlaena palustris, 5:135*n8*
Stenosemia aurita, 5:147*n1*
Stephania forsteri, 4:481*n9*
Sterculia
    foetida, 1:337*n14*, 3:73*n1*,
        4:137*n18*
    treubii, 3:77*n1*, 3:113*n4*,
        3:587*n3*, 4:321*n23*
Stercus squillarum, 5:509
Stickman, Olaf, 1:142, 1:143
Sticky Plum, 3:49*n10*
Sticky tree, 3:47, *3:48*, 3:49
Stictocardia campanulata, 4:477*n7*
Stinker tree, *1:332*, 1:333–1:334
Stipa tenacissima, 5:33*n7*
Stock Tree, 2:309–2:311
Stockfish Wood, 3:272*n36*, 3:501
Stoechas celebica, *5:554*, 5:555
Stokvis-hout, 3:502
Stone pearl, 5:426
Stone seed, *5:40*, 5:41
Stone-Breake, 5:140*n13*
Stramonia, 2:343
    Austriacam, 4:180
    indica, 4:177, *4:178*, 4:179–4:182
Strand-Nagelboom, *3:368*
Strawberry shrub
    black, 3:427, *3:428*, 3:429
    red, *3:424*, *3:425–3:426*
Strychnos
    barbata, 3:553*n3*
    colubrina, 2:179*n1*
    muricata, 2:179*n7*
    nux-vomica, 4:179*n17*
Stylophora digitata, 5:430*n8*
Styrax
    benzoin, 2:85*n6*, 2:87*n24*,
        2:304*n15*, 3:549*n14*, 3:559*n15*
    benzooin, 4:63*n36*

calamita, 2:281, 3:666, 3:667
liquida, 2:83, 2:597
officinalis, 2:281*n2*
Sualan, 1:257
Suavadi, 3:372
Suba gratsji, 5:279
Subergorgia mollis, 5:407*n1*
Subtiliter serrata, 3:507
Suca duca, 5:283
Sucan, 4:258
Succinum terrestre, 5:235
Succisa, 4:249
Succun tsjii, 3:590
Suda malam, 4:241
Sudu
    lapas, 3:293
    sudu, 3:329, *3:330*, 3:331–3:333,
        3:332, 5:581
    ticos, 5:582
Suelen, 3:187
Sugar
    apple, 1:399*n1*
    cane, 3:209, *4:92*, *4:93–4:97*,
        5:277*n10*
    palm, 3:258*n13*, 5:507*n4*
    plum, 1:340
    reed, 3:209
        bastard, 3:211
Suihumate Tree, 2:281
Sulamu taly, 4:515
Sulanking, 5:77
Sulassi, 4:205, *4:206*, 4:207
    ayer, 5:292, 5:293
        kitsjil, 5:291
    kitsjil, 4:213
    mihic, 4:207
    puti, 5:293
Sulica, 5:273
Suluk, 4:268
Sumanca, 4:432
Sumangi, 4:229
Sun plant, 5:72
Sunda Assa Tree, 2:134

Sune hahatu, 1:352

Sunflower, 4:247*n5*
  wild, 5:545*n4*

Suniha, 3:276
  furu, 3:271
  laut boubou, 3:273

Sunty, 4:55

Supa, 3:11, 3:13, 3:270

Suppliant plant, 5:211*n2*

Sura-nipa, 1:290

Suren Tree, 2:526, 2:527–2:528

Surenus, 2:526, 2:527–2:528, 2:528

Suri, 5:173

Suru madjangan, 3:333

Susanna flower, 5:227*n1*

Susehion, 2:56

Sussu, 5:343

Sussuela, 4:541
  esculenta, 4:539, *4:542*, 4:543–
    4:544

Sutjang, 3:270

Suur-kroop, 5:529

Suyn-cam Tree, 2:155

Swallowwort, 5:495*n4*
  Indian, 5:101*n2*

Swart-boom, geplekte, 2:424

Swarte Londt-boom, 3:113

Sweat-Weed, 3:217*n2*

Sweet Chinese Citrus, 2:168*n3*

Sweet flag, 4:323*n31*, 5:353*n3*

Sweet Lemon, 2:168*n6*

Sweet sop, 1:397*n12*, 1:399*n1*

Sweitzer-hosen, 4:192

Switches
  blood red, 5:403, *5:404*

Sycomorus
  chartaria, 3:37
  indica, 3:36

Symphonia, 4:168

Symphytum, 5:57*n1*, 5:59*n5*

Symplocos javanica, 2:605*n8*, 3:51*n6*,
  3:55*n7*, 3:57*n1*, 3:271*n19*,
  3:411*n5*, 5:95*n9*

Syngamia abruptalis, 4:205*n10*

Syra, 1:194, 1:257

Syringa
  alba, 3:555
  arabica, 3:558

Syzygium
  aqueum, 1:377*n1*
  aromaticum, 2:3*n1*, 2:4*n13*,
    2:543*n1*, 2:543*n3*, 2:548*n10*,
    3:161*n1*, 3:275*n19*, 4:183*n2*,
    5:208n11, 5:222*n12*, 5:439*n1*
  cumini, 1:385*n2*, 1:387*n1*,
    1:388*n6*, 3:141*n5*, 3:142*n12*
  jambos, 1:373*n13*
  javanica, 1:381*n14*
  malaccense, 1:369*n2*, 1:375*n1*,
    1:388*n7*, 3:121*n1*, 4:183*n2*
  stipularis, 1:383*n17*

T

Ta-iipa bonko, 1:325

Täa, 4:303

Taanboom de kleyne, 3:118

Taarki Dizaar, 1:256

Taba, 4:186

Tabaco, 4:149–4:150
  laut, 3:385
  utan, 5:107, 5:108, 5:111

Tabaculi, 5:122

Tabalissa, *5:13*, 5:15

Tabatico, 3:181
  Ake, 3:195
  java, 3:192
  nani, 3:205
  sammat, 3:189
  tuy, 3:189

Tabe, 4:315
  kling, 4:186
  lomboc, 4:186
  rawit, 4:186

Tabernaemontana, 2:372, 2:373,
  2:375
  capsicoides, 3:417*n1*
  divaricata, 2:194*n23*, 3:327*n1*
Tabernaemontana capsicoides,
  5:101*n3*
Table leaf, 4:27, *4:28*, 4:29–4:30
Tablecloth-Leaf tree, *3:80*, 3:81
Tabu rottang, *4:92*, 4:93*n1*
Tacamahaca Orientalis, 2:269
Tacaossa, 5:301
Tacari root, 3:523
Tacca, 5:298
Tacca
  besaar, 4:307
  culat, 4:307
  fungus, 4:305*n1*
  gunong, 4:312
  latifolia, 4:305, *4:306*, 4:307
  laut, *4:308*, 4:309
  leontopetaloides, 4:309*n1*, 4:358*n6*
  litorea, 4:301, *4:308*, 4:309
  montana, 4:301, *4:310*, 4:311–
    4:312
  palmata, 4:311*n1*
  phullifera, 4:301, 4:301*n1*, 4:305,
    4:306, 4:307
    agrestis, 4:312
  pinnatifida, 4:301*n1*, 4:305*n1*
  sativa, 4:301, 4:301*n1*, *4:302*,
    4:303–4:304, 4:358
  utan, 4:307
    kitsjil, 4:312
Taccari, 2:579, 4:91
Tackary, 5:9
  laut, 5:9
Tacki, 5:5
Taeru, 3:163
Taetsia fructicosa, 2:595*n8*, 3:313*n1*,
  3:382*n9*
Tafobo, 2:591
Tagalagnana, *5:26*, 5:27
  utan, *5:26*, 5:29–5:30, *5:32*, 5:35
Tagallo, 3:45
Tagera kattu, 4:239
Tagorera, 1:365

Taguerela, 1:365
Tahitian chestnut, 1:455*n1*
Tahom, 4:143
Tahu, 4:410
Tai udang, 5:509
Taipa, 1:327, 1:331
  baladun, 1:325
  pali pali, 1:325
Takanhola, 5:503
Takkada, 3:383
Takkan parran, 5:555
Tal, 1:269
Talagaha, 1:256
Talatamado, 1:256
Talauma, 2:309*n1*
  rumphii, 2:310*n12*
Talawruxiam, 1:256
Taleru tree, 3:163, *3:164*
Tali
  ajer, 3:577, 3:578
  api, *3:488*, 3:489–3:490
  aros, 5:369, 5:369*n1*, 5:371
  assam, 4:561
  ayer, 3:582
  babounji, 4:507
  bacompol, *3:584*, 3:585–3:586,
    3:587
  bubur, 3:471, *3:472*, 3:473, 3:599
  carbou, 3:503
  cuning, 3:535, 4:478
  dingin, 3:585
  gassogigi, 3:605
  gnemon, *3:484*, 3:485, 3:491
  hismu, 4:510
  itam daun kitsjil, 3:599
  morea, 3:579, *3:580*, 3:581–
    3:582, *3:583*
  payt, 3:542
  pissang, *3:602*, 3:603
  pule, 3:524
  racu ceru, *3:496*, 3:497
  sasawi, *3:592*, 3:593
  siri, 3:553
  tsjurli, 4:470
  ubi utan, 4:510
  utan, 3:477
Talipot tree, 1:300

Talissa, 1:466

Talissey, 3:96

Taljabo Sula Bassa, 2:565

Talla, 1:257, 1:495
tallan, 4:494, 4:506, 5:38

Tallahoho, 5:86

Tallan, 3:187, 4:491

Talpetti, 1:265

Talucca Tree, 2:*114*, 2:115–2:117

Talutu, 3:351

Taly
cupan, 4:503, *4:504, 4:505,* 4:506
cussu, *4:578,* 4:579
gorita, 4:556
labo, 4:556
rottu rottu, 4:556
toacca, 4:565
ubi, 4:506

Tamalapatra, 4:320

Tamalla, 3:187

Tamalussel Tree, 2:120

Tamara tonga, 1:363

Tamarind, 2:131, *2:132,* 2:133–
2:135, 2:380, 3:73, 3:83, 3:87,
3:90, 3:265, 3:266, 3:281, 3:565,
3:603, 3:635

Tamarindus, 2:131, *2:132,* 2:133–
2:135, 5:465
indica, 1:489*n3,* 2:99*n3,* 2:131*n1,*
3:266*n9,* 3:565*n2,* 3:603*n4,*
3:635*n1,* 4:267*n5,* 4:513*n17,*
5:152*n21,* 5:465*n3*

Tamarinjo, 4:140

Tamatta, 4:455, *4:456,* 4:457, 4:569

Tambar, 4:445

Tambido, 4:197

Tambu nuno, 3:625

Tame, 1:485

Tamhyo, 5:5

Tammalassin, 2:511

Tamme
Jamboezen, 1:372
krimp-bonen, 4:387

Tampu sosng, 3:311

Tamputi, 4:71

Tana sabrang, 4:377

Tanacetum vulgare, 3:591*n11*

Tanarius
alba, 3:116, 3:117–3:118
major, 2:546, 3:119, *3:120*
minor, 3:116, 3:117–3:118
rubra, 3:116, 3:117–3:118

Tanax, 1:344

Tancari, 4:91

Tando russa, 5:155

Tangkuli Tree, 2:123

Tangkulo, 1:506

Tanguli Ginten, 2:126

Tanidri, 4:461

Tanjong, 2:287, *2:288,* 2:289–2:290,
2:434
beach, 2:291, 2:292, 2:293
Laut, 2:291, 2:500
wild, 2:291, 2:292, 2:293

Tanjonus litorea, 2:291, 2:292, 2:293

Tankoë, 4:421

Tankschi, 3:336

Tanni-Paina, 1:225

Tanning tree
large, 3:119, *3:120*
small, *3:116,* 3:117–3:118

Tannini, 5:527

Tansey, 3:591*n11*

Tapanawa, 4:572
badiri, 4:577
kitsjil, 4:581, 4:582
parvifolia, 4:571
tairis, 4:581

Tapeinidium amboynense, 5:139*n4*

Tapeinochilos ananassae, 5:72*n15,*
5:559*n1*

Tapeussa, 5:301

Tapile, 3:190

Tappa rawa rawa, 1:325

Tappagadja, 4:477

Tappal inan, 4:477

Tappauch, 1:372

Taratti, 4:463, 5:322, 5:325, 5:327
    kitsjil, 5:326
    sajor, 5:326

Taraxacum officinale, 4:135*n7*

Tarenna buruensis, 2:507*n2*,
    2:637*n2*, 3:350*n8*

Taro plant, 4:285*n2*

Tarragon, 5:555*n3*

Tarron, 4:141, *4:142*, 4:143–4:147

Tartar plant, 4:83

Tartarski zeli, 4:83

Tarum, 2:59

Tat wel, 1:344

Tatal-iba, 3:507

Tataru mäa, 3:585

Tatepal, 3:209, *5:26*, 5:29–5:30

Tatexar, 3:211

Tatimanu, 1:256

Tatisjitu, 1:256

Tatta som, 5:548

Tatume aur, 3:407

Tarupele, 3:209

Taü, 4:401

Taunui, 5:60

Tavarcarre, 5:384, 5:390

Tavernaemontana capsicoides,
    4:564*n6*

Tawawe, 2:631

Tawe, 4:506

Tawera, 4:239

Taxotrophis ilicifolia, 2:517*n1*

Tay
    angin, 5:170, 5:341
    burong, 3:569, *3:570*, 3:571–
        3:572
    gigi, 3:61, 3:63, 3:572
    hông, 5:287, 5:288

Tayris, 4:455

Tea, 1:116

Tea plant, 5:86

Teba teba, 1:51, 3:199, 3:203

Tebe pussa, 5:277

Teben meccati, 4:86

Tecabbal, 5:301

Tecche, 4:186

Tecke
    laki laki, 5:5
    pantey, 5:5

Teckee, 5:4, 5:5, 5:19
    laut, 5:9

Tecomanthe amboinensis, 5:543*n5*

Tectaria crenata, 5:133*n5*

Tectona grandis, 2:463*n1*, 3:73*n5*

Tedjo, 2:96

Tehu lopu, 5:277

Tehuila, 5:126

Teley, 1:466

Telinga ticos, 5:243, 5:346

Telosma
    cordata, 5:571*n1*
    odoratissima, 5:571*n1*

Tema tennalla lanit, 4:19

Temantong, 4:351

Tembul, 4:320

Temu besar, 4:65*n3*

Tenarron, 4:143

Tenga, 1:215

Tenga Maran, 1:196

Tenna talile, 4:12

Tephrosia purpurea, 5:281*n1*

Terbang saccu, 5:150

Terebinthina, 5:284, 5:291

Teregam, 3:41

Terin
    3:191–3:194

Terin Hahuru
    3:195

Terminalia
    alba, *3:312*, 3:313–3:314
    angustifolia, 3:315, *3:316*, 3:317,
        3:461
    arborea, 3:520*n11*
    bellerica, 2:124*n15*

catappa, 1:465*n1*, 2:207*n2*,
  2:229*n4*, 2:268*n19*, 2:473*n6*,
  3:61*n4*, 3:95*n3*, 3:97*n2*,
  3:169*n2*, 3:431*n6*, 5:309*n4*
  chlorocarpa, 1:466*n6*
  macrocarpa, 1:465*n2*
  rhodocarpa, 1:465*n4*
  chebula, 2:124*n15*
  citrina, 2:124*n15*
  rubra, 3:312, 3:313–3:314
    silvestris, 5:541, 5:542
    sylvestris, 3:382, 3:461, 4:577
Ternstroemia robinsonii, 2:589*n5*,
  3:169*n1*
Terong, 4:169
Terre djangan, 3:501
Tetech Tree, 2:57
Tetipoteriba, 3:616
Tetracera
  boerlagei, 3:489*n2*
  scandens, 3:489*n1*
Tetraclinis articulata, 2:391*n9*
Tetrastigma, 4:515*n5*
Tette Palehu, 3:67
Tewer, 2:374
Tewu, 4:95
Thaumantias, 4:192
Theka, 2:465
Themeda
  arquens, 5:27*n1*
  giganta, 3:213*n10*, 5:29*n1*
Thenghio, 2:5, 2:7
Thespesia populnea, 2:339*n1*,
  2:346*n16*, 3:131*n4*
Thi Seê, 2:431
Thieve's nutmeg, 2:26, 2:35
Thimhio, 2:54, 2:55, 2:56
Thistle, 4:195*n3*
  blessed, 4:135*n5*
  milk, 4:261*n2*
  sow, 4:261*n2*
  spindle, 4:136*n15*
Thorn apple, 4:177*n2*
Thorny tree, 5:533, 5:534, 5:535

Thou thau, 4:474
Thouboytsi, 4:499
Thuarea involuta, 5:21*n4*, 5:21*n5*
Tick, 3:181
Tielapong, 3:192
Tiger flower, 5:184*n59*
Tigil kio, 4:261
Tigo, 5:86
Tihing, 3:188
  ampel, 3:205
Tikos Trec, 2:494, 2:495–2:497
Time Flower, 3:222, 3:223–3:224,
  3:230
Timon passir, 4:435
Timon tree, 3:172, 3:173
Timonius, 3:172, 3:173
  timon, 3:173*n1*
Tin Uyn, 2:379
Tinaspora crispa, 3:609*n1*
Tinat, 3:189
Tinder, Chinese, 4:203
Tingi lacca, 3:251
Tingulong tree, 5:561
Tinospora
  reticulata, 4:245*n11*
  rumphii, 5:571n3
Tintinga, 3:365
Tiri macke, 5:243
Tiru calli, 5:581
Tiru makir, 4:553
Titawey, 3:27
Tithymalus, 2:360
Titsjiu, 1:320, 5:359
Titti Laut, 2:473–2:474
Tittie Tree, 2:469, 2:470, 2:471
  beach, 2:473–2:474
Tittius, 2:469, 2:470, 2:471
  litorea, 2:473–2:474
Tjacca mabido, 3:546
Tjang, 3:270
Tjomu tuffa, 5:295

Tlattan cuya, 4:316

Tòa thau kiong, 4:69

Tobacco, 1:232*n11*, 1:233, 1:250,
  4:149–4:150
  wild, 5:108

Tobba toko, 3:225

Tobo Tobo, 3:45

Todda
  panna, 1:300, 1:319, 1:320
  vaddi, 4:272

Toeblaadjes, 4:485, 4:486

Toeplitzella regia, 5:389*n37*, 5:401*n1*

Tohong, 4:169
  canjii, 4:173

Toja, 4:303

Tola topolo, *5:13*, 5:15

Tolasion, 2:565

Tolassi, 4:207

Tomatl, 4:171, 4:455

Tomato, 4:171*n20*, 4:183*n3*,
  4:455*n1*
  strawberry, 5:115*n1*

Tommon
  ackar, 4:65–4:66
  badur, 4:61, 4:63, 4:66
  besaar, 4:65–4:66
  contsji, 4:66, 4:67, 4:71
  dingin, 4:66
  giring, 4:66, 4:69
  itam, 4:66, 4:69
  lawac, 4:65–4:66, 4:68, 4:69
  manga, 4:67
  pho, 4:66
  plant, *4:64*, 4:65–4:69
  puti, 4:39, 4:66, 4:68
  ties, 4:68
  tihing, 4:66
  utis, 4:66
  white, 4:66

Toncat langit, 4:19

Tonckat seytan, 3:215

Toncke, 2:601

Tondjo, 5:321, 5:326
  bado, 5:326
  puti, 5:326

talipu, 5:326

taratte, 5:326

Tonka langit, 4:466

Tonke Parampuan, 2:607

Tonokuko, 2:79

Too kocki tsjaptec, 5:495

Toona sureni, 2:527*n1*

Tooth rubber, 3:605

Toothpick plant, 4:494

Topikki, 3:38

Torch Tree, 2:553, 2:554
  large, 2:549, 2:550, 2:551, 2:552

Torminalis, 2:384, 2:595

Tornasol flower, 4:192

Toron, 4:169

Totarra Itam, 3:109

Totarre Palecong, 3:81

Tou leallun, 3:351

Touch-me-not, 4:269*n18*

Tournefort, 4:467

Tournefortia argentea, 3:385*n1*

Toxicaria minor, 5:309

Trachyphyllia geoffroyi, 5:437*n1*

Trachyspermum ammi, 4:219*n2*,
  5:77*n6*

Trandunja, 3:266

Trang dunja, 3:266, 3:282

Tree of Blind Eyes, *2:358*, 2:359–
  2:361, 2:362, 2:363
  uses, 2:365–2:367

Tree of Darkness, 2:561–2:562

Tree of Pointed Flowers, 2:287,
  2:288, 2:289–2:290

Tree pustules, *4:548*, 4:553

Tree Reed
  Cratium, 3:187–3:188
  Leleba, *3:180*, 3:181–3:185
  rough, 3:195, *3:196*, 3:197–3:198

Trema amboinensis, 1:491*n12*,
  2:571*n19*, 3:399*n1*, 5:509*n2*

Tremella marina, 5:331

Trichomanes indicum, 5:159, 5:160, 5:161

Trichosanthes
  cucumerina, 4:441n1
  trifolia, 4:453n4

Trifolium acetosum, 4:269

Trigonella foenum graecum,
  3:267n23, 4:63n41, 4:141n5,
  4:233n1, 4:411n2, 5:44n17,
  5:281n7

Triple flower, 5:214, 5:221–5:222

Tripneustus gratilla, 5:237n3

Tristis Tree, 2:287

Triticum
  repens, 5:19n2
  saracenicum, 4:115

Triumfetta bartramia, 4:113n3,
  5:113n1, 5:114n6

Trong
  andjing, 4:175
  brown, 4:169
  Chinese, 4:169n6
  East-Indian, 4:171
  gatta, 4:175
  mera, 4:169
  pra, 4:171, 4:175
  puti, 4:169
  sina, 4:171
  tamatte, 4:169
  tsjina, 4:169
  utan, 4:173, 4:176
  white, 4:169

Trongum
  agreste, 4:172, 4:173
    album, 4:171, 4:172, 4:175–
      4:176
    rubrum, 4:175n3
  hortense, 4:169, 4:170, 4:171
  pressum, 4:172, 4:175–4:176
  spinosum, 4:172, 4:173

Trumpet leaf, 4:27, 4:28, 4:29–4:30

Truna malam, 4:241

Tsada bangi, 4:126

Tsalan, 5:89

Tsampadaha, 3:4

Tsampe Tree, 2:296

Tsang, 1:279

Tsapitsou, 5:49, 5:50, 5:51, 5:52,
  5:53

Tschaupo, 4:83

Tschêen, 4:143

Tschéug, 3:25

Tschili, 4:183, 4:185, 4:186, 4:187,
  4:188, 4:189
  ayer, 4:183, 4:186
  besaar, 4:186
  cuning, 4:186
  mera, 4:186
  utan, 3:416, 3:417, 5:101

Tschilli
  abbal, 3:417

Tschina, 4:497–4:500

Tschintschau, 5:170, 5:171,
  5:171n19, 5:171n25

Tschira, 3:507

Tscho, 3:381

Tschongsidji, 3:84

Tschoo-hoae, 5:170

Tschui, 3:571

Tschyo, 4:6

Tsdina pelouke, 1:405

Tsili, 3:150

Tsja tick, 3:193

Tsjabe, 4:315
  utan, 3:547, 3:549
    tauwar, 3:550

Tsjaca
  maram, 1:345

Tsjacca
  magoli, 5:32, 5:36, 5:37–5:39

Tsjaccarbebe, 4:227, 4:228

Tsjad, 2:391

Tsjadaen, 4:258

Tsjaitau, 4:383

Tsjalimuri, 2:344

Tsjamanca, 4:432

Tsjamarkyen, 3:345

Tsjamarrha, 3:365

Tsjambu, 4:287

Tsjammo, 3:401, 3:445

Tsjampaâ, 2:84

Tsjampacca, 2:61, 2:62, 2:223,
    2:301, *2:302*, 2:303–2:305, 3:8,
    3:109, 3:134
    Gunong, 2:310–2:311
    mountain, 2:307, *2:308*, 2:310
    Utan, 2:310–2:311
    wild, *2:306*, 2:307

Tsjampada, 3:108

Tsjampadaha, *1:346*, 1:347–1:349,
    1:350*n1*, 3:107
    utan, 2:449

Tsjampadari boll, 5:237

Tsjampaga
    Babyan, 2:317

Tsjampangam, 3:272

Tsjampe, 3:266

Tsjampulong, 2:323

Tsjanghing, 2:355

Tsjankring, 2:355

Tsjappo tsjappo, 5:335

Tsjargay, 1:256

Tsjatick, 3:206

Tsjatsjisji, 5:539

Tsjatsjoran laut, 5:453

Tsjauw tsjeen, 3:513

Tsjaykoäe, 4:437

Tsjeck, 4:110

Tsjederecali, 3:331

Tsjekani, 1:226

Tsjeme tsjeme, 5:282

Tsjencke, 2:3–2:9
    Papua, 2:17, *2:18*, 2:19
    Radja, 2:17, *2:18*, 2:19
    Utan, 2:21

Tsjendana, 2:67
    Ackar, 2:69
    Djingi, 2:73, 2:75
    Mera, 2:75

Tsjendawan, 5:249

Tsjendikithoe, 3:568

Tsjenke
    laut, *3:368*, 3:369, 5:439
    utan, 3:161
        daun kitsjel, 3:401
    utan Kitsjil, 3:162

Tsjenti alibach, 4:193

Tsjeramela, 2:552

Tsjerekan, 4:561

Tsjeri nari nam pulli

Tsjero, 4:213

Tsjerou Ponna, 2:591

Tsjerri, 4:146

Tsjeru
    cansjava, 4:126
    kirganeli, 5:81
    parva, 5:87
    tardavel, 5:101

Tsjetsjang, 3:270

Tsjetti
    manduri, 3:267
    pu, 4:201

Tsjialiu Tree, 2:137

Tsjidji wood, 2:74

Tsjikiu, 5:10

Tsjili, 3:295, 5:326

Tsjincka hoe, 3:254

Tsjinckhi, 3:96

Tsjindaga, 3:435

Tsjinga tsjinga, 4:470

Tsjinkahu, 3:251

Tsjinkin tree, *5:578*, 5:579

Tsjintschaw javanense, 5:547

Tsjit tsjit, 4:268

Tsjitick, 3:199

Tsjitsoansom, 5:548

Tsjiu Tsjat, 2:393

Tsjiulang, 3:514, 5:537

Tsjo, 3:193

Tsjombor, 3:219

Tsjonckor, 4:48, *4:70*, 4:73–4:74,
    5:471

Tsjonkodo kir, 3:427

Tsjonkor, 3:320

Tsjonloo, 5:588

Tsjontsjo, 3:567

Tsjori valli, 4:516

Tsjoro mamma, 4:234

Tsjorro, 3:34

Tsjotsjo, 4:528

Tsjudemaram, 3:300

Tsjug-Jek, 2:383, 2:384

Tsjui camtsio, 5:293

Tsjukiet, 2:383, 2:384

Tsjulang, 3:516

Tsjupa, 4:149

Tsuggilida, 3:331

Ttamarindus indica, 3:83*n2*

Tuba
    baccifera, 3:526, 3:527–3:529
    flava, 3:527, 3:533, *3:534*, 3:535,
        4:478
    parampuan, 3:539, 3:540
    radicum, 3:527, *3:530*, 3:531–
        3:532
        nigra, 3:532
    siliquosa, 3:539, *3:540*
    sterilis, 3:141, 4:559

Tuba Root
    white, *3:530*, 3:531–3:532

Tuba shrub
    berries-bearing, *3:526*, 3:527–
        3:529
    pod-bearing, 3:539, *3:540*
    yellow, 3:533, *3:534*, 3:535

Tubastrea, 5:358*n13*

Tuber
    regium, 4:497, 5:233–5:235,
        5:235
    sampadarium, 5:237
    terrae, 4:497

Tuberosa, 4:242

Tuberose, 2:303*n10*

Tubi poccol, 5:86

Tubipora musica, 5:423*n1*, 5:449*n2*

Tubu, 4:95
    sala, 4:86, 5:29, 5:29*n3*
    tubu, 5:277, *5:554*, 5:559

Tubu Tubu Waät Kitsjil, 2:627

Tubularia, 5:424

Tuca dussa, 4:19

Tudong Laut, 2:639

Tuha
    ackar, *3:530*, 3:531–3:532
    mette, 3:532

Tuhe tuttunu, 4:559

Tuigkhiaa, 3:188

Tulip
    Javanese, 4:273, *4:274*, 4:275
    water, 5:321*n7*

Tulipa javana, 5:299*n3*

Tulussi, 4:209

Tundjo tedjalor, 5:326

Tundju langit, 5:295

Tundo langit, 4:19

Tunkat padang, 5:555

Tupac, 3:100

Tupha-indi, 1:372

Turbinaria
    conoides, 5:345*n7*
    ornata, 5:345*n2*, 5:345*n4*
    reniformis, 5:358*n13*

Turi, 1:489, *1:490*, 1:491–1:492,
    1:493
    mera, 1:492

Turia, 1:491

Turiaa, 1:485

Turmeric, 1:353*n12*

Turn-Leaf Tree, 2:596, 2:597–
    2:598

Turnip, 4:373
    East-Indian, 4:379

Turpentine plant, *5:284*, 5:291

Tussiaria, 5:65

Tutu Massoniha, 2:582, 2:591

Tutukeku, 3:381

Tuvera, 5:301

Ty, 4:303

Tylautassi, 2:511

Tylosurus, 3:442*n*5

Typarehu, 3:67

Typha latifolia, 1:256*n*8, 1:350*n*5

Typhonium divaricatum, 4:295*n*1, 4:569*n*3

Tythymalus, 2:369, 2:370, 3:329

Tzeda, 5:426, 5:427

U

Ua

ahun tayn, 3:647

boa, 5:426, 5:427

cawa, 3:659

ela, 3:647

helite, 3:652

huay, 3:656

mette, 3:643

Ubat

caloway, 3:71

dingin, 3:329, 3:331

guna, 4:272, 5:67, 5:80

Lemon, 2:203

lute lute, 4:165

massin, 3:276

matta, 5:291

murit, 5:72

papeda, 3:163, 3:402, 3:420, 3:421

radja, 4:498

rambu, 3:315

red, 5:95

rede, 3:119

rhambu, 4:477, 4:487

roston, 3:273

saguer, 1:277, 2:505, 2:511, 5:259, 5:561

swangi, 3:545, 5:51, 5:56, 5:57–5:58, 5:59–5:61, 5:259

Thuni, 2:203

Tuni, 1:277

Ubi

ahua, 4:351

alea, 4:339

ayam, 4:336

boaja, 4:343

bulu, 4:337

calabessa, 4:354

castela, 4:370

daun soccun, 4:339

elan, 4:351

gorita, 4:365, 4:366, 4:367

kepas, 4:339

lacki lacki, 4:336

pariaman, 4:358

radja, 5:233, 5:241

sussu, 4:373, *4:374*, 4:375

tangan, 4:339, 4:357

taun taun, 4:345, *4:346*, 4:358

torana, 4:345

ular, 4:343, 4:344, 4:345, *4:440*, 4:441

utan, 4:357, 4:361, 4:366

Ubium, 4:334, 4:335–4:338

anguinum, 4:335, *4:342*, 4:343–4:344

anniversarium, 4:335, 4:345, *4:346*

bulbosum, 4:335

caseolare, 4:335

digitatum, 4:335

majus, 4:339, *4:340*, 4:341

minus, 4:339

draconum, *4:342*, 4:343–4:344

fructiferum, 4:503

mammosum, 4:373, *4:374*, 4:375

nummularium, 4:494, 4:503, 4:504, 4:505, 4:506

ovale, 4:350, 4:351

polypoides, *4:364*, 4:365–4:367, 4:494, 4:506

polypoides nigrum, 3:350, 5:298

pomiferum, 4:335, 4:347, *4:348*, 4:349, 4:506, 4:507

quinque folium, *4:356*, 4:357–4:358

sativum, 4:336

silvestre, 4:290, 4:358, 4:359, *4:360*, 4:361–4:363

vulgare, 4:335, 4:336

Ubo ubo roriha, 3:219

Uccu manjoffu, 5:19

Ucke Puti, 2:353

Ucken Ewan, 2:449

Ucken lacki lacki, 3:417

Udani, 3:143, 3:591, 3:593

Uga uga, 5:277

Uge
  baja, 4:159
  bira, 4:459
  jabba, 4:511, 4:569
  kelo, 1:485

Uha utan, 5:543

Uhe
  talla, 4:366
  waleä, 4:366

Uhel, 3:119

Uhuta, 5:231

Uin kion, 4:60

Ujay lako, 3:275

Uken ewan, 5:487

Ula pala pala, 3:659

Ulal, 1:357

Ulang ulang, 3:16, 3:19, 3:30,
  4:523n13, 4:569, 5:559n12

Ulassie Tree, 2:479, 2:480, 2:481

Ulassium, 2:481

Ulat, 5:241, 5:245
  anahan, 5:237
  ela, 5:239
  hatu, 5:233
  kitrib kitrib, 5:243

Ulerum, 1:438

Ulet Tree, 2:516, 2:517

Ulex europaeus, 5:377n2

Ulibita, 4:361

Ulihelawan, 2:499

Ulipockol, 2:561

Ulit, 1:439

Ulit Helawan, 2:502

Ulua mentalis, 5:304n22

Ululon, 1:438

Ulvaria musaria, 3:603n2

Umarea, 3:113

Umbilicus
  rupestris, 4:527n2
  veneris, 4:527

Umbraculi figura, 5:242

Umbraculum maris, 2:639

Una nepa, 5:262

Unam, 4:63

Unapur, 4:300

Uncaria
  cordata, 3:577n1
  gambir, 2:379n8
  longiflora, 3:123n5, 3:573n1,
    4:209n4
  setiloba, 1:234n30, 3:573n4

Undia, 4:389

Undor undor, 5:49, 5:50, 5:51,
  5:52, 5:53

Une bin, 3:215

Unguis odoratus, 3:523, 3:659, 5:520

Unin, 4:60
  mackay, 4:48, 4:77
  packey, 4:47

Unipan, 5:262

Unmatta caja, 4:179

Unti
  java, 1:417
  unti, 1:417

Unting, 4:6

Untj Untj, 2:627

Unyn
  mahu, 4:67
  tita, 4:61, 4:67
  tommo, 4:67

Upas, 2:401n16, 3:125
  bidji, 3:273, 3:275
  camarungi, 3:273
  Daty, 2:474
  Lacki Lacki, 2:382, 2:383, 2:385
  Puti, 2:192
  puti, 3:294

Upessa
  abbal, 3:321
  hatu, 3:323

Ure rerel, 4:11

Urena labata, 4:177n3, 5:113n1,
  5:113n3, 5:114n6, 5:289n3

Uri
    bengala, 4:153
    mangala, 4:153
    wonin, 5:165
Urit, 1:439
    ehu, 1:439
Urtica
    decumana, 5:88, 5:89–5:90,
        5:89n1
    molucca mortua, 5:88, 5:91
Uru
    pickal, 5:233
    weru, 5:487
Ury maressu, 1:420
Usnea, 5:167n1, 5:169n3, 5:341n4
Ussaye, 2:513
Usse
    Crawo Tree, 2:155
    Ela Bal Tree, 2:152
    Malumeit Tree, 2:153, 2:154
    Mettin Tree, 2:165
    wary ewan, 3:510
Ussi
    Helawan Tree, 2:159
    Lapia Tree, 2:151
    Layn Tree, 2:177
    Wale Tree, 2:143
Utar, 1:357
Uteren, 1:438
Utetheisa lotrix, 4:232n7
Utta
    aman, 5:49, 5:50, 5:51, 5:52, 5:53
    baën, 4:159
    baja, 4:159
    bria, 4:459
    hahuru, 5:130
    hatu, 3:368, 4:528, 5:233
    lala, 4:568, 4:569
    lau cana, 3:660
    mahu, 3:595, 4:511, 5:115
    manu, 3:87, 3:279, 4:237
    matte, 3:598, 4:543
    meyt, 4:545
    mina, 3:612
    niwel, 1:313, 5:298
    nuer, 1:313
    papeyite, 5:136
    pela, 1:499
    puti, 5:233
    renut, 4:459
    soä, 1:477
    tela, 1:499
    tila, 1:499
    tohu, 3:373
    waa, 5:129
        lapia lari, 5:136
    waha lauila, 5:133
    wanant, 3:599
    weyer, 3:423
Utta-manu Tree, 2:123
Utte
    aule, 3:184
    lauit, 3:189
    onitu, 3:184
    wani, 3:195
    wannat, 3:187
Uva crispa Americana, 4:173
Uvas
    d'aljofare, 4:516
    de boy, 4:517
    dos bugios, 4:519
Uwer, 1:377
Uyn-som, 5:548

V

Vallaris glabra, 3:555n2
Valvulae, 3:133
Vanda
    crassiloba, 5:203n14
    furva, 5:197n1
    saxatilis, 5:203n14
Vandopsis lissochiloides, 1:496n6,
    5:193n1
Vanilla planofolia, 5:203
Varinga, 3:2, 3:3–3:9, 3:436, 3:587
    funicularis, 3:16, 3:607
    grossularia, 3:405
    latifolia, 3:5, 3:31
    microphyllos, 2:56, 2:65, 3:387,
        3:511, 4:175
    parvifolia, 3:24, 3:494
    repens, 3:11
Varnish Tree, 2:391, 2:392, 2:393–
    2:395

Varu Ranu, 2:332

Vegetable
black, 4:237n8
chicken, 4:237n7
Dutch, 4:131n3
pulse, 4:336n10
sea, 5:170n11
white, 5:240n4
wood, 5:240n3

Velvet dock, 5:545n4

Ventilago maderaspatana, 3:471

Venus hair, *5:144,* 5:145

Venus' slippers, 5:179n3

Verangenes, 4:171

Veratrum
album, 4:59n7, 5:299n4
nigrum, 5:548n21

Verbascum thapsus, 5:513n3

Verbena, 5:49
bontii, 5:53
indica, 5:53, 5:577
rubra, 5:574, 5:577

Veronica, 4:529n2, 4:533, 4:535,
5:291, 5:291n2
cinerea, 5:129n9

Verrucella, 5:375n2
stellata, 5:403n1

Vertifolia, 2:596, 2:597–2:598, 5:481

Vesicaria, 5:115n2

Vessel Reed, 3:191–3:194

Vetch, 3:23, 3:249, 3:565, 4:389
African, 3:567n7

Vicia, 4:389n4
africana, 3:567
faba, 4:387n2

Vidara, 2:172, 2:173, 3:605
agrestis, 2:183, 2:184, 2:185–
2:186
littorea, 2:176, 2:177–2:178,
2:179, 2:183
puti, 2:174–2:175
utan, 2:186

Vidimaram, 3:49

Vidoricum
domesticum, 1:463
silvestre, 1:463, 3:105, *3:106,*
3:107–3:108

Vidorik tree, 1:461, *1:462,* 1:463

Vigna
cylindrica, 4:401n1
marina, 4:415n1, 4:485n6
radiata, 2:389n16, 4:405n1
sinensis, 3:137n3, 4:380n7,
4:383n1, 5:199n15
unguiculata, 4:383n1, 4:401n1

Villebrunea, 3:423n1
rubescens, 3:41n6

Villuru, 3:476

Viminalis, 3:470, 3:471

Vinca
minor, 4:419n1
pervinca, 5:57

Vinho de Palma, 1:194, 1:257

Vintaro, 2:370

Vinum
cos, 4:151
palmeum, 1:194, 1:257

Viola tricolor, 3:563, 3:563n4

Violaria cormandelica, 2:309

Viper's grass, 1:133, 4:367n15

Virgae sanguinae, 5:401, 5:403,
5:404

Virgin's herb, 4:395, *4:396,* 4:397,
4:419n1

Virgularia juncea, 2:406n14,
5:422n12, 5:457n6

Viscum, 5:173, 5:182n49
album, 1:224n9, 5:180
amboinicum, 3:569, *3:570,* 3:571–
3:572, 3:615
indica, 5:184

Vitex
agnus castus, 3:255n1, 3:255n2,
3:257, 3:373n2, 4:126n20
cofassus, 1:344n12, 1:392n22,
2:204n13, 2:451n1, 2:517n3,
2:559n5, 3:57n3, 3:513n6,
5:231n7
timorensis, 2:451n3

moluccana, 2:469*n1*
negundo, 3:259*n1*
paniculata, 3:259*n1*
parviflora, 2:451*n3*
rotundifolia, 3:255*n1*, 4:409*n2*
trifolia, 1:279*n34*, 1:408*n3*,
   1:424*n10*, 2:13*n20*, 2:106*n14*,
   2:451*n6*, 3:337*n10*, 4:137*n21*,
   4:182*n41*, 4:389*n3*, 5:109*n15*
Vitis
  alba, 4:569
    indica, 4:511, *4:512*, *4:513*
  arachnoidea, 4:519*n1*
  geniculata, 4:510*n13*
  nigra, 4:449, 4:515
  quadricornuta, 4:507*n6*
  trifolia, 4:515*n1*
Vlas-dodden, 4:121, 4:121*n13*
Vleesch-kruyd, 4:207
Vliegende duive, 5:187
Vlotgras, 5:337, 5:337*n7*
Vogel-nesjes, 5:343
Volo Condan, 2:332
Volvaria volvacea, 5:239*n1*, 5:239*n3*
Vuur-vare, 5:131
Vuur-wortelen, 4:521

W

Waal mendya, 1:345
Waan semane, 3:205
Wabba, 2:370
Wabbo Meyt, 2:385
Waboa, 5:426
Waccat, 4:358, 5:193, 5:195, 5:211
  bessi, 2:631
Wacun ranu, 3:453
Wadapu, 4:250
Waecor, 4:312
Wafelsteen, 5:437
Waffle stone, 5:437*n6*
Walan, 3:169
Walat, 1:283

Wale, 4:459
  bissi, 4:541, 4:549, 4:551
  makel, 4:579
  pinan, 4:572, 4:575
Wali
  ahuun, 3:585
  ammelaun, 3:553
  hulan, 3:535
    hakulu, *3:536*, *3:537*
  maluhu laun maun, 3:581
  maua, 3:603
  mette, 3:603, 3:608, 3:637
  puti, 3:593
  puti lohaha, 3:617
  soa, 3:485
Walissiva, 3:598
Wall-reed, 3:199, *3:200–3:201*,
  3:202
Wallacea, 1:1
Wallflower, 3:269
Walo, 3:671
Waloc, 4:425
Wanake, 3:199
Wanbaysen, 3:234
Wangcudu, 3:51
Wangi, 4:329
  malacco, 4:261
    tali, 4:263
  wangi maiho, 5:83
Wani, 1:334, 5:262
Wanlecu, 5:187
Wannat, 3:187
Wanyer, 3:103
Warama cussu, 5:71
Waran, 3:169
Waranderck, 2:344
Wari
  ahuun, 3:587
  ammo, 3:554
  hismu, 4:510
  hulan, 4:478
  lavian, 3:471
  lottu lottu, 4:507
  metten, 3:503, 3:599
  tua, 4:565

unenar, 4:263
wari puti, 3:219
waruhu, 3:581

Warihokor, 2:536

Waringa, 1:347, 2:603, 4:6, 4:7,
　4:553, 5:105, 5:182*n48*, *5:208*,
　*5:240*n6, 5:481

Waringin, 2:417*n12*, 2:611, 3:16,
　3:108, 3:508, 3:517, 3:571, 3:572,
　3:579, 3:586, 3:607, 3:612
creeping, *3:10, 3:11*, 3:12, *3:13*
daun bezaar, 3:5
daun kitsjil, 3:24
Indian, 3:8
malatta, 3:11
Pitsjaar, *3:26*, 3:27, *3:28–3:29*,
　3:30–3:31
reddish, 3:11
small-leaved, *3:22*, 3:23–3:25
tsjampadaha, 3:5

Warisahu, 4:479

Wariuhi, 4:506

Wash-wood, 2:351*n18*

Water
cinnamom, 5:293
greens, 5:335, *5:336*
jamboezen, 1:377
lettuce, 5:333–5:334
licorice, 5:293
Meloenen, 4:431
melon, 4:429
　Indian, *4:430*, 4:431–4:434
mint, 5:292, 5:293
pompoenen, 4:423
ritsjes, 4:183
rope, 3:577*n3*
salade, 5:333
stone, 5:449
tulpe, 5:321
wood, *3:422*, 3:423

Waterkwaat, 3:528

Watersteen, 5:449

Watery shrub
female, *3:354*, 3:355–3:356
male, 3:351, *3:352*, 3:353

Watta, 2:617
Buhulawan, *2:628*, 2:629
Cuban, *2:640*, 2:641
Cumban, 2:623
Ewan, 2:508
Lopu, 2:601
Mahina, 2:605, 2:607, *2:628*,
　2:629
Tube Tube, 2:639

Wawakkal, 3:103

Wawaran, 3:243

Waweru, 4:401

Way way, 1:327, 1:331

Waylo hittuta, 4:489

Weasel root, 5:523*n6*

Weddor, 4:485
lau mau, 4:487

Wedelia biflora, 2:625*n4*, 3:350*n13*,
　4:201*n17*, 4:263*n5*, 4:469*n1*,
　4:469*n6*, 4:472*n24*, 5:71*n3*

Wedule, 4:485

Week-boomen, 3:84

Weinmannia fraxinea, 3:163*n1*,
　3:401*n5*

Wellat, 3:405

Welluang, 5:541

Welluayt abbal, 3:315

Welluga, 3:314

Wentletrap plant, 5:275, *5:276*,
　5:277, 5:559*n1*

Weri, *5:32*, 5:33–5:34, 5:304

Werihassa, *5:26*, 5:29–5:30

Wersingan, 5:541

Werusissi, 3:314

Wewean, 5:335

Wheat
grass-like, 4:116
Turkish, 4:115

White Tree, 2:103, *2:104*, 2:105–
　2:107
myrtle, *2:108*, 2:109
small, *2:108*, 2:109–2:110

White vegetable tree, *1:494*, 1:495–
　1:496, 1:497, *1:498*, *1:499*

Whore Flower Tree, 2:303

Wida dari, 3:427

Widjin, 4:119, 4:120, 4:121
  puti, 4:122

Widorik, 1:463
  utan, 1:463, 3:108
    pandjang, 3:105
  wild, 3:105, *3:106*, 3:107–3:108

Wier, 5:346, 5:353

Wild Bitangor, 2:577

Wild Cassia Tree, *2:128*, 2:129

Wild Drumstick Tree, 2:127, *2:128*,
  2:129–2:130

Wild red border shrub, 5:541,
  5:542

Wilde Vygen van Oost-Indiën,
  3:34

Willow, 3:47, 3:91, 3:159, 3:293,
  3:368, 5:469

Windbollen, 5:10

Wine
  apple, 1:361*n3*
  wood, 5:507

Winteranus cortex, 2:92

Witte, 4:401
  kraalboompjes, 5:430

Wocka, 1:247
  wocka, 1:265

Woit, *3:606*, 3:607–3:608
  besar, 3:607
  kuri, 3:608
  lain ella, 3:608
  laun mäun, 3:608

Wolf's claws, *5:164*, 5:165

Wolf's Milk, 2:360

Wolfbane, 4:296*n3*

Woman's Leaf, 3:371

Wonder Tree, 3:336*n6*

Wonderboom, 3:336

Wondrous Quis-Qualis shrub,
  3:588, 3:589–3:591

Wonga landac, 5:511

Woo, 3:38

Wood for flies, 5:497

Wood leaf, 5:503

Wood of the Lords, 2:585

Wood plant, 5:171

Woodsorrel, 4:229, *4:230*
  Indian, 1:31

Wool-bearing tree, 1:503

Woraka, 4:55

Worm cushion, 4:523

Wormwood, 2:179, 2:367,
  3:591*n11*, 5:519, 5:519*n4*

Wort, 4:59*n9*

Wortel-boom, 2:601

Wotcl, 5:113

Wrapper plant, *5:218*, 5:219–5:220

Wrightia javanica, 2:129*n5*,
  4:513*n16*

Wuka utan, 1:269

X

Ximenia americana, 2:162*n7*,
  2:177*n1*, 3:605*n4*

Xiphium indicum, 4:79

Xiuhquilitl pizahoac, 4:145

Xylocarpus
  granatum, 2:582*n4*
  moluccensis, 1:277*n15*, 2:581*n1*
  rumphii, 2:582*n9*

Xylophyllos ceramica, *5:502*, 5:503–
  5:504

Y

Yal
  Amin, 2:250
  Halat, 2:226
  Mette, 2:242

Yalanassoy Tree, 2:237

Yam, 5:34*n17*, 5:225*n19*

Yaniri Tree, 2:245

Yar issi apasua, 5:163

Yaro mahu, 2:271

Yasesel Tree, 2:245

Yellow Flower, *3:278*, 3:279

Yellow Wood, 3:501*n24*
Ygo ygo, 5:80
Yhur, 1:266, 1:269
Young Woman's Tree, 3:391
Youngia japonica, 5:65*n6*
Yu-kam, 5:465, 5:465*n7*
Yucca, 4:304
    American, 4:358
Ywel, 4:358
Yzerhout, 2:431
    bastart, 2:449
    gemeen, 2:440

Z

Zabid, 4:136
Zaga, 3:85
    frutex, 3:83
    pohon, 3:84
Zagu Plant, 3:230*n10*
Zajor bajang, 5:279
Zalacca, 3:661, *3:662*, 3:663
    edulis, 1:334*n11*
Zalack shrub, 3:661, *3:662*, 3:663
Zanthoxylum, 2:195*n1*
Zea mays, 4:115*n7*
Zedoar, black, 4:39
Zedoaria officinarum, 4:67
Zedoria, 5:5
Zee-basta, *5:450*, 5:451
Zee-boonen, 3:476
Zee-kroost, 5:349
Zeekraal, 5:411
Zeenagelen, 5:439
Zeewajer, 5:376
Zehneria indica, 4:537*n3*
Zenka, 4:354

Zenserlin, 4:119
Zergjants-klooten, 3:663
Zeruba, 4:39
Zerumbet, 4:64, 4:65–4:69, 5:5
    claviculatum, 4:70, 4:71
Zeuxine amboinensis, 5:175*n1*
Ziet maar eenen dag, 5:309
Zingiber, 5:219*n5*
    aromaticum, 5:34*n19*
    caninum, 4:48, 4:186
    cassumunar, 3:332*n20*, 3:427*n10*,
        4:63*n42*, 5:151*n18*, 5:555*n12*,
        5:559*n5*
    fuscum officinarum, 4:48
    minus sive gramineum, 4:50,
        4:55–4:56
    officinale, 3:275*n18*, 3:314*n4*,
        3:621*n5*, 3:626*n14*, 3:660*n13*,
        4:48*n5*, 4:49*n3*, 4:49*n4*,
        4:55*n1*, 4:445*n7*, 5:5*n14*,
        5:15*n9*, 5:51*n17*, 5:126*n13*,
        5:167*n3*, 5:261*n5*, 5:313*n10*,
        5:430*n9*
    purpureum, 3:96*n9*, 3:294*n8*,
        4:47*n1*, 4:84*n9*, 5:80*n19*,
        5:222*n11*, 5:261*n4*, 5:268*n6*
    zerumbet, 2:403*n32*, 2:405*n1*,
        3:294*n9*, 3:427*n9*, 3:427*n14*,
        4:37*n1*, 4:39*n3*, 4:41*n3*,
        4:48*n4*, 4:57*n2*, 4:84*n10*,
        4:323*n30*, 5:34*n19*, 5:223*n8*,
        5:310*n11*, 5:471*n19*, 5:559*n4*
Zizyphus, 2:174, 2:175, 2:185, 3:250
    jujuba, 2:138*n11*, 2:173*n1*,
        3:250*n7*
Zong zay, 5:479
Zoraco, 2:496
Zuikerwerk, 5:435
Zwarte boontjes, 4:409

Colophon

The text of this book is set in 9.5/11 point Janson,
designed by Miklós Tótfalusi Kis.
The notes are set in 6.5/8 point Giacomo Light,
designed by James Montalbano.

The plates were scanned by the Missouri Botanical Garden
from an original copy of *The Ambonese Herbal* in their collection.

All files for production were prepared on Macintosh computers.

It was printed and bound in Ann Arbor, Michigan
by Sheridan Books, Inc.

The paper is 60 pound House Natural Smooth B18.

Production was managed by Maureen Noonan.

It was edited by George Scott and was indexed by Carol Inskip.

It was designed by Charles Nix and Alexandra Zsigmond.